Getting to Grand Prairie

―∭―

One Hundred Londoners and their Quest for Land in Frontier Illinois

Getting to Grand Prairie

—◊—

One Hundred Londoners and their Quest for Land in Frontier Illinois

Karen Cord Taylor

Weathergage Press
115 Town Line Road
Francestown, NH 03043
www.gettingtograndprairie.com

Cover image "In Concert" ©Larry Kanfer, www.kanfer.com

Copyright © 2015 by Karen Cord Taylor
All rights reserved.
Printed in the United States of America

ISBN: 9780990439509
ISBN: 099043950X

For Russell V. Puzey, Genevieve Carter, Ralph E. Houser, and Ruth Clipson Anderson, who got her wish.

Acknowledgments

I am indebted to many people in both America and England who helped me put together the story of the Grand Prairie English. First are Sara Cast, director of the Catlin Historical Society, and her assistants, Virginia Wallen and the late Ruth Martin. I also want to thank Tammi Addington, Findlay, Ohio; Roberta Allen, Danville Public Library, Danville, Illinois; Arthur Anderson, Eagle Lake, Texas; David Axford, Stanford in the Vale, Oxfordshire, United Kingdom; Lynne and Dick Barnes, Urbana, Illinois; Tom Belton and Dolores Bedinger Belton, Indianola, Illinois; Sally Bruce-Gardyne, Aswardby, Spilsby, Lincolnshire, United Kingdom; Eve Carson, Indianapolis, Indiana; Richard A. Chrisman, Bloomington, Illinois; Elizabeth Sutherland Clark, South Bend, Indiana; Janet Cornelius, PhD, Penfield, Illinois; Jack Cumbow, Collinsville, Illinois; Sam Davol, New York, New York; Herb Depke, Cary, North Carolina; Tom Dye, Macomb, Illinois; Fred Faulstich, Faulstich Printing Company, Danville, Illinois; Margaret Sutherland Fraser, Lossiemouth, Scotland; William Furry, Illinois State Historical Society, Springfield, Illinois; Chad Hanna, Berkshire Family History Society, Reading, Berkshire, United Kingdom; Bill Hensold, Danville, Illinois; Ginger Hern, West Challow, Oxfordshire, United Kingdom; Sally Hinkle, Boston, Massachusetts;

Mona Church Hunter, Allerton, Illinois; Sue James, Bodleian Library, Oxford University, United Kingdom; Jenny Marie Johnson, University of Illinois Library, Urbana, Illinois; Lee Pritchard Kershner, Las Cruces, New Mexico; Mark Learnard, Georgetown, Illinois; Peter and Nancy Maitland, London, England; Suzanne Kinder Miller, Estes Park, Colorado; Don H. Morrison, Monee, Illinois; Ben and Jennifer Newell, Savoy, Illinois; Robert W. O'Hara, British National Archives, Kew, Richmond, Surrey, United Kingdom; Guy Puzey, Edinburgh, Scotland; Dave Reynolds, Champaign, Illinois; Vicky Sax, Watertown, Massachusetts; Billy Sherrill, Valparaiso, Indiana; William Spencer, British National Archives, Kew, Richmond, Surrey, United Kingdom; Scott Steward, New England Historic Genealogical Society, Boston, Massachusetts; Ronald Story, PhD, South Deerfield, Massachusetts; Jorie Walters, Kankakee County Museum, Kankakee, Illinois; Lawrence Ward, Stanford in the Vale, Oxfordshire, United Kingdom; and Courtney Wood, Evanston, Illinois.

These people helped with research, loaned me family treasures, pointed out mistakes, connected me with other helpful people, and read the manuscript for errors and omissions. Some even put me up for a night or two. Any mistakes are my own.

Such books as this are only a beginning. In the best of circumstances, they spark interest in a period and a place. They help ferret out additional information that adds to the story. I hope others will add to the narrative of the people who settled one small part of Illinois.

"People make history by passing on gossip, saving old records, and by naming rivers, mountains, and children."
Laurel Thatcher Ulrich, *Well-Behaved Women Seldom Make History*

"But property, property sticks, and property, property grows."
Alfred, Lord Tennyson, "Northern Farmer: New Style"

Contents

Preface · xiii

1 July 6, 1862. Catlin Township,
   Vermilion County, Illinois ·1
2 The Long View · 7
3 A Green and Pleasant Land · · · · · · · · · · · · · · · · · ·15
4 A Rift Among Puzeys ·25
5 Another Land of Lincoln ·33
6 Lincolnshire Ties ·39
7 Lighting London ·49
8 Ladies and Gentlemen ·61
9 Another Kind of Life ·69
10 Gathering for the Grand Prairie · · · · · · · · · · · · · · ·81
11 A Girl's Life in London ·93
12 Tossed at Sea ·99
13 Claiming the Prairie · 119
14 The Special Case of Will Clipson · · · · · · · · · · · · · 133
15 Ambivalent Citizens · 145
16 Lords of the Soil · 159
17 The Americans' War · 171
18 A Legacy of Gooseberries · · · · · · · · · · · · · · · · · · 185

Epilogue ·201
Maps and Illustrations ·203
A Guide to the Grand Prairie English Families · · · · · · · · 211
Selected Bibliography ·219
Index ·227

xi

Preface

Abraham Lincoln *is* Illinois history for many people. Sometimes events in Chicago add to the state's narrative. Because Illinois has been identified so strongly with an iconic, martyred president and an important metropolis, other stories have been less explored.

This book fills in one of the gaps. It is the story of several large families and their friends, more distant relations, and acquaintances—about a hundred people in all—who immigrated in the mid-1800s to Illinois' Grand Prairie, a colloquial name for the region extending south from a growing Chicago along the Indiana border. Their particular destination was southern Vermilion County.

These families came from England, which provided 18 percent of Illinois' foreign settlers during this time. They were lured not by streets paved with gold but by the land that Illinois possessed so much of. They had observed that landowners in England led lives of leisure, high status, and impressive civic responsibility. They wanted that kind of life for themselves, and they sought it in America's Midwest.

It took the heads of these families until they were in their forties to make the move. Except for Matilda Clipson and Richard Puzey, most of the family heads died after only a decade or so in their new home. So much of their story takes place in England.

xiii

Getting to Grand Prairie

These people did not achieve the fame of Abraham Lincoln, but their stories are nonetheless fascinating. The contrast between Lincoln and the Grand Prairie English is the difference between the *Mona Lisa* and genre paintings. The *Mona Lisa* is renowned, but genre paintings, with their details of everyday life, portray stories worth investigating. They also provide the context for the history of the better-known.

One of these families was my own. I grew up in Catlin Township in Vermilion County. The stories my father told were puzzling. He claimed his great-grandfather had been wealthy and influential in England and had grown more so through his Illinois land. I was skeptical. Why would such a family leave their homeland for the Illinois prairie when they weren't starving or fleeing religious persecution, as many nineteenth-century immigrants were? Why would they decide to be farmers? Why would they choose Catlin, Illinois, eight hundred miles inland from the port at which they landed, over other places in America?

I decided to seek answers to these questions. As I gathered information, new questions arose. I couldn't figure out why they had come to Vermilion County, so I investigated the hunch that my family must have known someone who had come before. Many immigrants to America have chosen their destinations that way—a practice known as chain migration.

Gradually I realized my family had not come alone but had coordinated plans with dozens of London friends and relatives. The first clue was the 1850 United States census in Vermilion County. Certain surnames kept cropping up— the same names as many of my elementary and high school friends. The members of those families had all been born in England. Curiously many of the names were clustered on a couple pages, meaning those families lived near one another. I discovered they had bought adjacent land within

xiv

a few years' span. English census records showed some of the families had lived near one another in London. Ship records showed when my friends' families had emigrated from England. Some came about the same time or even on the same ship as my English ancestors.

Through the Catlin Historical Society, I learned that three of the English families of that period had stories written down. I found that family stories have kernels of truth, even if they are distorted. I found they are mostly distorted.

Within these family stories, however, was original material—letters, diaries, documents, and pictures—revealing the immigrants' motives and the consequences of their move. Several of the families had visited one another in London. It became obvious that these Londoners had planned and saved for their trips together. Over a period of about six years, they headed for a small part of the Grand Prairie where the boundaries of what became Georgetown, Catlin, and Carroll townships meet. As I interviewed people, I found they knew parts of their family's story. Some realized their ancestors had come with another family. But no one understood the extent of the connections.

Some of my best friends in high school were Suzanne Kinder, Gary Reynolds, and Bonnie Jones. I discovered that Suzanne's great-great-grandparents and mine had been friends in London in the 1840s. Gary, Bonnie, and I shared great-great-great grandparents. We were distant cousins. In all the years we saw one another every day, not one of us learned of these connections. How could such connections have been forgotten?

It turned out what I had set out to do wasn't unique. I was, in fact, part of a large, growing movement of amateur historians trying to discover who their ancestors were, what they did, why they did it, and what it meant in their time.

More than any other reason for the new interest in family history is that it can finally be done. A wealth of primary sources—census data, passenger lists, regiment lists, wills, and birth and death records—becomes newly available almost weekly on the Internet. The information is international in scope and searchable by names, places, and dates. This makes possible an activity that previous generations, including professional historians, gave up in frustration.

Verifying stories passed down in families is a risky business. Many stories have only shreds of truth in them. "Facts" fondly passed through the generations turn out to be exaggerations, products of the imagination, or simply false. For example, the histories made much of the immigrants' high rank in England. It turned out, however, that with a few exceptions, the immigrants came from a class of skilled workers with enough money in their pockets to buy land when they came.

One person cautioned me about revealing the truth when the false stories were more entertaining and elevated the stature of the ancestors. In most cases, however, the verified accounts turned out more interesting than the less reliable tales the families passed down. Some stories I heard were scandalous. Others were preposterous. If I couldn't verify an account, I didn't include it.

I have not included many photographs of the people in this narrative. Existing photographs are mostly of poor reproductive quality, or the subjects of the photographs can't be identified with certainty. A caution to readers: many people in this story have the same name. Sarahs, Williams, Jameses, and Elizas abound. To help keep members of families straight, refer to the list of family members and the dates of their arrivals in America in the back of the book.

Tracing the past sometimes reveals mundane but surprising realizations. As I learned about Henry Jones's

Preface

birthday gooseberry pie and Richard Puzey's long row of gooseberry bushes, I realized why my grandmother, a daughter of an English immigrant, had planted the bush from which I picked those round, smooth, puckery fruits I loved as a child as much as Henry and Richard apparently did. Once, as an adult, I showed an American friend living in the Netherlands how to make a pie from the green and red gooseberries in her garden. She had not even known they were edible. I can't easily find gooseberries now because I live in New England, where their cultivation is discouraged. The gooseberry and its relative, the currant, are thought (perhaps erroneously) to carry a disease that kills the region's majestic white pines.

Some of the readers of this book are likely to be descendants of the Grand Prairie English. They now live all over the United States. One grows rice in Texas. Another is a retired school principal in Colorado. A third is a teenage cellist in New York City. There are teachers, doctors, ministers, lawyers, coaches, a travel agent, bankers, computer specialists, soccer players, a glass artist, accountants, writers, at least two hospital executives, and, of course, farmers. They live in every state in the Union, and some still live in Illinois.

The stories of the Grand Prairie English are unique, but they share a universal thread. These immigrants had certain dissatisfactions with their circumstances in the old country, they heard the United States would offer them a better life, and they possessed a sense of purpose and adventure. They are why we are here.

Karen Cord Taylor
Boston, Massachusetts
April, 2015

1

July 6, 1862. Catlin Township, Vermilion County, Illinois

On the morning of the last day of his life, Will Clipson crawled out from under the bedclothes, went to the window, and looked out over his fields. They were dripping with dew.

The day would be a scorcher. The cattle with their thick coats must already be uncomfortable. The heat, however, was certainly good for the wheat. That patch was pure gold and almost ready to harvest.

Another field was still only partially broken. Will didn't know how he would manage with all his boys gone. Breaking the sod was heavy work, and the heat and humidity slowed them down. The equipment broke easily, and Will was not good at fixing things. He dreaded the rattlesnakes that glided through the grass fleeing the plow, and he feared the tornadoes that sucked into their maws anything in their way. He hated the prairie flies that flew into his face as he worked. The summer's heat made his head ache, but the fierce winter cold was just as intolerable. The Grand Prairie was as flat as the pastoral Lincolnshire landscape he had left behind but shared none of its peacefulness or civility. *This place isn't fit for human habitation,* he thought, not for the first time.

Or enjoyment either. There was nothing to look at. No canals to break the endless land into manageable parts. No hustle, no bustle, no London. He missed his old life, and he didn't share his friends' conviction that owning all the land beneath his window was worth the effort. It was hard to get to town, and when he got there, there was little to see except the muddy streets. The Americans were so devoted to their nonconformist churches that there were few amusements. No betting. No good pubs.

And his family was disintegrating. Over the years he'd watched his first wife and fourteen of his children die. Now with this American war, which was not of his doing, he could lose his three grown sons.

Billie, his oldest boy, had already gone to Decatur to volunteer and was serving in Company A of the Twenty-First Illinois Regiment somewhere in Missouri. The younger boys were still around. They were probably already at breakfast downstairs, but they wouldn't be home for long. Jack had signed up with Company D of the 125th Illinois Infantry and would leave in August. James would turn eighteen in a couple weeks and was determined to sign up on his birthday.

Will hadn't been able to stop Billie or Jack. Billie thought joining the infantry would make him more of an American, and besides, all his friends were enlisting. Jack didn't need a reason. He'd always been a hothead.

James was cautious and more responsible, but that responsibility meant he felt pressure to sign up from the other men and boys. He also knew his father was still able enough to run the farm.

July 6, 1862. Catlin Township, Vermilion County, Illinois

He couldn't save Billie or Jack, but he could save James.

Will closed the bedroom door quietly since Matilda was still asleep. He went downstairs but avoided the kitchen. He walked outside and down the lane to the barn. The rope was where it always was—on a nail just inside the door. He tossed the rope over the tie beam and secured it with a loop.

Then William Henry Clipson, late of London and aged fifty-six, tied the rope around his neck and hanged himself.

—∞—

At least that's a story that came down in family lore. Will's descendants believed he killed himself so James would be the only man left to continue farming. This would make it impossible for him to enlist. After his father's death, James did stay on the farm.

Some descendants said they had heard Will committed suicide because one of his sons had been killed in the Civil War, and he couldn't bear it. That is not true. His sons survived the war, although not happily. Another part of the family was of the opinion that his sons were waiting at the train station to join their regiments when a neighbor rode up to tell them of their father's death. That dramatic scene didn't happen either, since none of the sons was leaving on that emotional day.

No official death record exists for William Henry Clipson. Perhaps it burned in the 1872 fire that destroyed the Vermilion County Courthouse. If an obituary was written, no one kept it. No one kept a record of who found Will's body.

Family members in later generations said they heard Will was sorry he'd left England. Some said he feared what they termed a "British mafia" that was out to get him. If he

3

Getting to Grand Prairie

left a suicide note, no one kept it or let it be known. The usually voluble Henry Jones, Will's close friend, left no letter describing his dismay at what Will had done. It's possible, though, that sorrow over Will's death hastened Henry's, since Henry lived only until November of that same year.

The other English who had moved to this small part of Illinois' Grand Prairie about the same time as Will were mum on the subject of his suicide. No one recorded the grief, anger, or regrets his family and friends must have suffered after they found him. If they did, those writings have been lost.

Will wasn't laid to rest in the English cemetery that Henry Jones had established—an important exception, since all the other English immigrants were interred there. Instead he was buried in the nearby township of Georgetown. Later, when his wife, Matilda, died, his body was moved to Oakridge, the cemetery in Catlin Township that welcomed everyone, no matter where they had come from. An 1889 collection of biographies of important men in Vermilion County described Will's success in creating a prosperous farm. It was silent, however, on the circumstances surrounding his death.

Yet every one of Will's descendants heard some version of the suicide story and talked about it for generations afterward. The suicide had such an impact that his descendants forgot to ask why Will was in that place at that time. Like most Americans, they knew the English had settled in America in the seventeenth and eighteenth centuries. But flooded with the nineteenth century's dramatic stories of desperate Irish, Italian, Eastern European, and Scandinavian immigrants, they and many others forgot the stories of the thousands of English immigrants who settled mostly in what is now called the American Midwest

4

July 6, 1862. Catlin Township, Vermilion County, Illinois

and whose assimilation must have been easier than that of other nationalities since they spoke the same language, and their conditions were generally not as dire. That's not to say, though, that the English adjusted easily to leaving family, friends, and the familiarity of their island's landscape.

A hundred fifty years later, Will's descendants and others in the community had lost the story of how he and dozens of his friends and relatives had endured the treacherous waters of the Atlantic Ocean and the Ohio River to seek wealth and status in a sparsely inhabited part of the prairie on the eastern edge of central Illinois that still had no official name or permanent boundary. They had lost the nineteenth-century prairie names based on soil types, topography, and history that had evoked for the immigrants such promise. The immigrants' indiscretions and imperfections, which for some had served as their reasons for immigration, were buried under layers of ennobling tales told by prairie descendants whose desire for virtue outweighed their interest in the far more interesting truth.

This, then, is that story of how and why about one hundred English men, women, and children, none of whom was fleeing the typical nineteenth-century emigrant scourges of starvation or religious persecution, together left their homes in and around London and settled among—or overwhelmed—the small population of American-born settlers who had moved to a particular area in the young state of Illinois.

2

The Long View

Tom and Dee Belton's Illinois farm lies in a landscape similar to the one Will Clipson would have viewed. It is just down the road from the land on which the English clustered when they arrived in the Grand Prairie in the mid-1800s. Both Beltons are descended from several of the English immigrants of that time.

The Beltons' house sits comfortably at the end of a long driveway on a low moraine left by a glacier thousands of years ago in the southern part of present-day Vermilion County. Beginning in the mid-nineteenth century, experts advised prairie farmers to plant tall, thorny hedgerows of Osage orange trees as natural fences. This would keep cattle in (or out) and prevent the wind from whipping up the soil and blowing it away. During the last part of the nineteenth century and about half the twentieth century, the hedge-rows changed the landscape that the early settlers would have encountered.

Over the last forty years, however, the hedgerows have gradually disappeared. Farmers who no longer keep cattle wanted the forty feet of acreage under the mature trees for their crops, and they ripped them out. So the Beltons' house overlooks uninterrupted fields of corn, soybeans, and sometimes a wheat field or two rising and falling into the horizon.

Although the farmers of the 1850s would have grown more wheat and no soybeans, they would recognize the setting. When a thunderstorm—or a tornado—is predicted, modern-day farmers can see it coming for a long time before it actually hits, just as those nineteenth-century settlers would have done. It is as hot and humid in the summer now as it was then, but there is one relief from the early aggravations. The rattlesnakes are gone.

The prairie sod is now broken, most effectively after the Civil War by John Deere's self-cleaning steel plow. Once again, though, it is left untilled, groomed by wide, complicated, folding chisel plows pulled by powerful tractors that point themselves down a row with the same GPS technology that can send a smart bomb toward a target fifteen miles away. The tractors' cabs are air-conditioned and outfitted with cup holders and stereo systems, making the present-day work more comfortable than it was for the early farmers.

As a boy Tom Belton worked alongside his father and then went to the University of Illinois thirty miles away in Urbana. He earned a degree in agriculture, helping him negotiate the planting, cultivating, fertilizing, and harvesting matters that his English forebears had had to learn on the job. He now farms about a thousand acres, some of it his land and some owned by others. His operation is typical for this part of Illinois. The land here is still in the hands of individuals and not large agribusinesses.

The Beltons run a family farm. It was partly inherited from relatives who also farmed it, but none of their sons is a farmer. Ben Belton is in billboard advertising sales in North Hollywood, California. Zach is a management

The Long View

consultant specializing in university hospitals. He lives in Chicago. Their third son, Will, is also in Chicago and works as a computer analyst with the Bank of America. Tom plans to turn the farming over to a niece and her husband when he retires in a few years.

The Beltons are an attractive, well-educated couple. They are students of American history, especially Illinois history. They participate in field trips sponsored by the local historical societies, and Dee serves on the board of one of them.

Their English ancestors would have known what we now call the American Midwest, especially Illinois, by its prairies rather than its administrative boundaries, since those boundaries were new and in flux.

The Grand Prairie, on which the Beltons' farm lies, was the largest of more than a hundred named Illinois prairies. It covered about a quarter of the state, stretching south from Chicago over an ancient lake bed about a hundred miles along the Indiana border and west toward Decatur. History, soil, and the types of vegetation growing on them differentiated this and other prairies. Rivers or stands of timber divided them. Buckeye Prairie was named after the Ohio settlers who made it their home. Walnut Prairie was named after the trees growing along its edges. English Prairie in southern Illinois was the site of the community of Albion, where a group of contentious English men and women who wrote extensively about their experiences settled there in the 1820s. Froggy Prairie and Big Mound Prairie were so named for obvious reasons. Goose Nest Prairie was the first home of Abraham Lincoln's family after they moved to Illinois from Indiana. The origin of some prairie names, such as Looking Glass Prairie, the tiny Gun Prairie, and even Lost Prairie, are lost.

9

Recently many groups have studied the soils, flora, and fauna of the native prairies. They have restored prairie tracts throughout the state where visitors can take walks and enjoy wild indigo, alliums, coneflowers, and grasses as well as the creatures that live among them. Less attention has been paid to the historic names, and few prairie aficionados are aware that in the first half of the nineteenth century prairie names were commonly used because the counties and townships were not yet firmly established.

The Northwest Territory, or "the old Northwest," of which Illinois is a part, was established by Congress in 1787. Surveyors began measuring the Illinois territory in 1804. They established baselines and prime meridians and gradually divided the land into townships, ranges, and sections. Once those boundaries were established, land could be bought and sold, and a buyer would know what he or she was getting.

The French had been the first Europeans to explore the Illinois territory. They were looking for furs and salt. They found salt on the Vermilion River near the area where the Grand Prairie English eventually settled. The river, and later the county, were named for the red clay along its banks. Salt was important for flavoring and as the principal means of preserving meat before refrigeration. If settlers were going to prosper in the Grand Prairie, they had to have salt for their food and as a commodity for trading or selling.

Illinois become a state in 1818, and the first permanent white settlers in present-day Vermilion County arrived at the Vermilion River in 1819, drawn by the promise of the salt works. At first they stayed in the woods that lined the river and its tributaries. They believed the prairies were infertile, since only grass grew there. The newcomers also

The Long View

correctly suspected the often wet and stagnant prairies carried disease.

They congregated at Butler's Point, named after the first settler, the Vermonter James Butler, where the forest intruded into the prairie. Soon after they congregated at places called Yankee Point, Quaker Point, and Brooks Point. These were also named after those first settlers—immigrants mostly from New England, Ohio, Indiana, and Kentucky. In 1826 Vermilion County came into being, covering what is now several other counties to the north and west.

By the 1830s, only a few Indian villages remained in this part of the Grand Prairie. Various tribes, including the Kickapoo, Piankeshaw, and Potawatomi, had ranged over this land in the eighteenth century, and some had established temporary villages. By 1834, however, there were not enough tribal members left to burn the prairie, an ancient undertaking that sustained the grasses. Those tribes that remained along the Illinois and Indiana border were summarily removed to the Kansas territory in 1838 in a march now called the Trail of Death. The name offers some idea of the condition the tribal members were in as they were herded through the area later to be inhabited by the Grand Prairie English.

The Beltons knew some of the English immigrants' story. Will Clipson's suicide was a dramatic incident involving the English in this part of Illinois, but he was only one of many adjusting or not to their new lives. Other English families related by blood or friendship, including the Puzeys, the Joneses, the Bentleys, and the Churches, had arrived in Vermilion County shortly before Will. Other immigrants, mostly single men related to earlier arrivals, would come later in the 1850s.

11

Getting to Grand Prairie

The Beltons were both descended from the Puzey family. In 1847 Richard Puzey was the first of the English settlers to migrate to this section of Grand Prairie. Although still a frontier, the region was gradually becoming more developed. He stayed with a family named Sandusky while he bought land and built a rudimentary house. The house was wood-framed, rather than made of logs as earlier houses were. Richard learned that schools had been started, and two years after he came he saw the Georgetown Seminary's two-story brick school building rise in Georgetown, the township adjacent to the land on which he had settled. Begun by the Methodists, it was incorporated into the public school system in the 1860s.

Soon another Methodist school, Vermilion Seminary, began operating in Danville, the largest town in the county. Not to be outdone, the Presbyterians also started a school, Union Seminary, while the Quakers already had one.

Farmers were still raising flax and sheep, and their wives and daughters were still spinning and weaving their own cloth. Boot and shoe makers still went from house to house and used a farmer's own animals' skins to make their products. These practices would change rapidly in the 1850s as ready-made goods began to appear.

There were still dangers, though. In a history of Vermilion County, Lottie Jones told the story of a girl whose mother and father had left her alone in the winter of 1850 in their house on the prairie:

After the hired man had gone upstairs to bed, the girl imagined she heard a faint sound; she ran to the door and threw it open. As the door was flung open their faithful shepherd dog bounded in. He was closely followed by a number of wolves who

12

The Long View

were chasing him and almost had caught him. They stopped when the light from the open door fell upon them. The girl hastily closed the door and shutting them out, shut the dog within.

A snowstorm arrived, wrote Lottie Jones, and when the girl looked out the window she saw the eyes of two wolves. They finally left in the morning when the snow stopped.

—⚏—

Dee Belton was a descendant of the first of the English immigrants, Richard Puzey. Tom was descended from Richard's brother Joseph, four of whose sons came to Vermilion County. So Dee and Tom shared the same great-great-great-grandfather in Berkshire, England.

Tom wasn't entirely English, though. "My dad's parents came over in steerage from Ireland," he explained. Other ancestors, the Sanduskys, were Polish. Members of this family had taken in the immigrant Richard Puzey while he got settled. Sandusky was an Americanization of Sodowsky—a surname said to have connections with nobility. Owners of vast tracts of Illinois land with which they speculated, the Sanduskys played a critical role in convincing the English emigrants to set their sights on Vermilion County.

"My Polish ancestors probably felt they had more cachet than the rest of my ancestors," he said. "But that didn't filter down to us since my great-great-grandmother was the daughter of Abraham Sandusky. She eloped with a plasterer and was disowned."

Despite a typically American mixing of ancestry for Tom, the Beltons enjoyed an English legacy. They were married at Fairview, the small country church still in the

Getting to Grand Prairie

business of worship that Richard Puzey and Will Clipson's widow, Matilda, had founded after the Civil War. The Beltons have pictures of their English ancestors and several books about the Berkshire villages from which they came. These were written by a family member whose Puzey forebears had remained in England. They also have a charming watercolor of Tom and Dee's shared great-great-great-grandfather's farm in West Challow in what is now Oxfordshire.

The Beltons had been to the Puzeys' English villages and met their distant cousins whose ancestors had remained in England. They were both vague, however, about the connections that other immigrant families might have had with theirs prior to their mid-1800s settlement in central Illinois.

3

A Green and Pleasant Land

By the time they decided to settle on the American prairie, Will Clipson and most of his immigrant friends were Londoners through and through. But some had started out on the farms of rural England.

The Puzeys originated in Berkshire. The farmer James Puzey—the father of six immigrants to America, the grandfather of many more, and the great-great-great-grandfather of Dee and Tom Belton—was relatively prosperous, owning land in the parish of West Challow. Nevertheless, in the 1830s, James's two older sons, John and James, left respectively for Ontario and New York State, popular destinations in that decade.

In 1847 James's fifth son, Richard, left too. Instead of following his older brothers, Richard headed for Illinois, which, unlike the eastern United States and Canada, still had plenty of available land. He was the first of his friends and family to do so. Fourteen members of the Puzey family and several friends and distant relatives followed him to the Grand Prairie over the next ten years.

The Puzey name is an old one. It was conferred on the Norman settlement of Pesei, which is now the hamlet and civil parish of Pusey. The letters "ey," as in the ancient Berkshire village and parish names of Pusey, Goosey, Charney, and Childrey, signify an island in the old Saxon language. In the case of Puzey, it refers to the "island of peas." Swamps surrounding islands of raised land characterized ancient Berkshire. Legumes were an early crop. Over the centuries the swamps were drained and turned into the tillable fields where sheep now graze or crops are planted.

Puzey can be spelled many ways. Pizzey, Pezey and Pewzey are only a few variations. As early as 1254, a Henry de Pesey was recorded as living in Berkshire, and in 1579 a John Pusey from Berkshire registered at Oxford University.

The name has a certain status in the region since it is connected with the eighteenth-century Pusey House, a grand manor with a lovely park built in the parish of Pusey by a family of landowning Puseys who died out. The house passed to the French Protestant Bouverie family, distant relatives who took the name Pusey as a condition of inheriting the house. Pusey House in Pusey should not be confused with Pusey House at Oxford University, which was also named after the family. In 2010 Pusey House in Pusey was put on the market for £27 million by its then-owner, a descendant of the family that started the Cadbury chocolate company in 1824. Angered by the way Kraft Foods had handled Cadbury since buying the historic enterprise, she reportedly intended to use the money from such a sale to start a rival chocolate company. The Grand Prairie Puzeys, however, are no longer connected to the family that owned the manor, if they ever were.

Historic Berkshire was divided from Oxfordshire, to the north, by the Thames River. The county lies to the west of London, beyond Heathrow Airport. Berkshire is now posh, moneyed, and filled with important visitor

A Green and Pleasant Land

sites. Windsor Castle is in Berkshire as is Legoland. Prince William's wife, Catherine Middleton, grew up there.

Berkshire, pronounced BARK-shur, was settled pre-historically by the Atrebates, a Celtic tribe from what is now Belgium and northern France. The Atrebates were accomplished traders, plying between the continent and southern England. The four hundred-year Roman occupation of Britain eclipsed their influence. In the 1870s someone discovered the foundations of a Roman building with a good view in West Challow. The building included a heating system, a well, and a chalk floor.

After the Romans left, the Saxons came from central Europe and settled in Berkshire and other parts of southern England. The name of the River Ock, a small tributary of the Thames that flows past Stanford in the Vale (one of the Grand Prairie Puzey family villages), comes from the Saxon word "ehoc," which means "salmon."

In the early eleventh century, a Puzey assisted the Danes in achieving temporary success over the Saxons in Berkshire, according to a story a Grand Prairie Puzey descendant learned when he was in England. He recapitulated it for his American family:

> During the time in which the Danes were trying to take over the Anglo-Saxons, the Danes were camped at Pewsey. A fellow, also named Pewsey, disguised himself as a shepherd and got inside the camp of the Anglo-Saxons. Their army was there in Northern [*sic*] England inside of a great big swamp. Pewsey was supposed to blow a horn when the Anglo-Saxons were going to attack. Of course, he blew the horn, and the Danes under Canute got ready and attacked the Anglo-Saxons when they arrived. With this

forewarning the Danes won the battle. So the king gave Pewsey all the land where the horn could be heard. [He ordered a man to] go up on a hill [and] blow the horn, and Pewsey was given all the land wherever it was heard. They had people stationed around to listen for the horn. Then Pewsey was awarded the bull's horn that was used. This created the Puzey estate, which was established in 1014 and was not broken up until 1929.

The writer of this version of the tale, the late Russell V. Puzey, was a great-grandson of the immigrant Richard Puzey. Russell saw the horn in the 1950s at the Victoria and Albert Museum in London, where it still resides. It has been embellished with metal and an inscription that translates to, "I, King Canute, gave William Pusey this horn to hold thy land." No evidence exists about whether this early Pusey, whose land defined this portion of Berkshire, was related the Grand Prairie Puzeys.

In any case Canute and his heirs ruled England until 1042 when Saxon kings were reinstated. In 1066 William the Conqueror swept into England at Hastings and overwhelmed southern England. Soon Saxon rule was no more.

The French kings dissolved into the Plantagenets and so on. These changes sometimes affected Berkshire farmers and sometimes not. Berkshire was heavily contested during the English Civil War, but probably the biggest change in the county began in the late 1790s. Berkshire underwent enclosure, which must have heightened the Puzeys' sense of the importance of land ownership.

Enclosure, which transformed land held in common into privately held tracts, had been going on for hundreds of years all over England. According to tradition since the

A Green and Pleasant Land

Middle Ages, the king owned all the land in England. He granted parcels of his land to his lords in exchange for their loyalty, protection, and certain other services. A lord then presided over what was called a manor, which contained the demesne, the farm that generated income for the lord. (The word "domain" derives from "demesne.") An upper hierarchy of peasants held the remainder of land in the manor. They worked what was essentially their own land, but they and laborers with no land also had to work on the lord's demesne for the privilege of living on his manor. Large portions of the manor, though, were shared for grazing and raising common crops.

By the late 1700s, this manorial system had largely died out in England. Agricultural laborers were paid in cash, and farmers paid rent to the person who was literally the lord of the land, or the landlord. After the rent was paid, they were pretty much left alone to do with their land what they pleased. This included passing it on to their children. But vestiges of the old system remained. This included the existence of much common land, which was often poorly maintained. Sharing it was thought to be an inefficient way of farming. The practice of enclosure was designed to distribute the land to individuals, thus remedying some of these problems and bringing farming up to date.

Due to acts of Parliament, Stanford in the Vale and West Challow, two of the villages in which the Grand Prairie Puzeys lived, quickened the pace of enclosure at the end of the eighteenth century. This was just as the future emigrants were being born. The large open spaces of common land that the local people shared for livestock grazing and other farm activities were divided into individually owned allotments, surrounded by hedges or ditches.

19

Getting to Grand Prairie

For some farmers this was a disaster. Rich landowners often received land the small farmers had historically used in common. The new owners would enclose it for their own benefit, leaving the small farmers with no means of support. Sometimes, though, the small farmers, especially those who believed they had chances at allotments, supported the practice.

The Beltons' great-great-great-grandfather, James Puzey, was a farmer with small holdings who profited from enclosure. The Enclosure Award for West Challow, dated 1803, shows that a house in the village and more than thirty-nine acres of formerly common land were awarded to "James Pizzey, also Puzey, a one-time church warden at the little church [St. Lawrence] close by." At the time West Challow's population was 185 persons. Over the next few years, James Puzey also received common land in other parts of the parish.

Other Grand Prairie English surnames such as Tarrant and Lloyd also appear as recipients of common land from the practice of enclosure.

The villages that were home to the Puzeys who immigrated to the Grand Prairie, however, are no longer in Berkshire. County reorganization transferred them to Oxfordshire in 1974.

—◊◊◊—

Childrey, West Challow, and Stanford in the Vale, villages in which the Puzey family lived, lie within five miles of one another in the Vale of the White Horse, a broad plain bordered by ranges of low hills. Atop one of those ranges lies the Ridgeway, a long path said to be the oldest road in

Britain. From the path walkers or riders on horseback can enjoy what is truly England's green and pleasant land.

On one side of the ridge topped by the ancient roadway is a remarkable site—the prehistoric, 374-foot-long figure of the Uffington White Horse carved into a chalk hill. A hard scouring every seven years traditionally kept the chalk exposed and bright. The future Grand Prairie immigrants surely would have participated in the maintenance of this figure. According to historians, local people kept it up well into the 1800s. English Heritage, the organization that oversees Stonehenge, is now responsible for caring for the figure. During World War II it was covered to protect it from German pilots bent on destruction and hoping to use it to find their way. Local residents say the best view of the White Horse is from Dragon Hill, said to be where the mythical St. George slew the dragon.

The vale is what many Americans picture when imagining England. The villages with their lush greenery and well-tended gardens have small stone churches surrounded by toppled gravestones. Fresh thatch covered with chicken wire forms the roofs of many cottages, although some are now covered with slate, and some villages have undergone recent development. Hawthorn hedges line the narrow roads. English daisies dot the grass. Pheasants strut across the roads, and hares poke up their heads, seemingly to watch the infrequent traffic go by. Many fields, which were farmed as small holdings in the mid-1800s, are now almost as broad as fields in Illinois. Instead of corn and soybeans, though, a common crop is the rapeseed plant with its vivid yellow blossoms.

The rough-plaster two-story house in West Challow that James Puzey was allotted through enclosure was where Richard and Sophia, two of his children who immigrated to the Grand Prairie, grew up. It still stands, pretty and well maintained. It lies alongside Childrey Brook and is conveniently near St. Lawrence, the parish church where James Puzey served as a warden. The oldest part of the house dates from at least the seventeenth century but might be earlier. When the Puzeys were living there, it was surrounded by orchards.

A house across the street has a sign reading "Fairview." Fairview indeed has a fair view, as it looks over meadows and fields that back up to the little village. Fairview was the name given to the area of the Grand Prairie where Richard Puzey and the other English immigrants bought land. It was also the name of the Methodist church where Dee and Tom Belton married. It's possible Richard chose the name because he remembered his neighbor's house fondly.

In rural England, however, Fairview meant something besides a pleasant landscape. It would have overlooked an open space, a village green, large enough to accommodate seasonal fairs or weekly markets before these villages had permanent shops.

That's exactly where Fairview is in West Challow. An old map marks the intersection of narrow roads between the former Puzey house and Fairview as the church green, a space large enough to hold fairs. Originally Fairview was two houses with separate front entrances. In earlier eras animals lived with people on the ground floor of the house, or the animals lived on the ground floor with the people above. A former owner converted one of Fairview's front doors into a window and removed one of the staircases to the upper floor. Two matching rooms with fireplaces form the front of the house. Plaster covers the original brick and timber walls, and exposed beams raise the low ceilings to a

A Green and Pleasant Land

more comfortable height. Someone built an addition onto the back that contains a kitchen on the ground floor and a bedroom and baths above.

An investment banker and his wife now occupy Fairview. He takes the train into London's financial district every day, as do many residents of these Oxfordshire villages. In the twenty-first century Oxfordshire is as posh as Berkshire. Neither county was well-off when the Grand Prairie Puzeys lived there. One local writer called Stanford in the Vale "tumble-down" in 1859. That was about ten years after the Puzeys left.

The Puzeys' villages now are pretty and prosperous. They offer few clues as to why so many would leave them. But other records—censuses, remembrances, birth and death certificates, and books—help tell the story of people who, were they still alive, might wish more facts had remained hidden.

4

A Rift Among Puzeys

Why would the Puzeys leave England for the American prairie when they already owned good farmland in Berkshire? The answer might partially lie with one man—the Beltons' shared great-great-great-grandfather.

James Puzey was lucky in many ways. For one, he lived to a ripe old age. This was not the case with his siblings. James was the youngest of the thirteen children of John Puzey of the village of Childrey and Mary Seymour from Ashbury. Parish records show James was born on May 23, 1765, in Childrey and christened in June of that year. He married Ann Tarrant, twenty-five, from Stanford in the Vale about three miles away when he was twenty-two. The ceremony took place at the St. Denys Parish Church in Stanford next door to the Tarrant manor house. Ten of James and Ann's children survived into adulthood in an era when children's lives were precarious. James's own parents hadn't been so lucky. When his father, John, died in the mid-1790s, James inherited his father's farm, since his older brothers had died.

By 1803 James had added to his father's land the thirty-nine acres he received in West Challow and other parcels he acquired through enclosure, so he was a farmer with small holdings but some means. In 1820 he leased Eastfield Farm, which totaled almost two hundred fifty acres, in Stanford in the Vale for fourteen years, and the family moved there. His relative prosperity enabled him to pay for apprenticeships in baking, brickmaking and drapery for his younger sons. If he still owned the enclosure land, presumably he would have passed it on to his oldest surviving son. There is no evidence, however, he did so and no clue about what happened to the land he owned outright or leased. According to Lawrence Ward, archivist and librarian at the Stanford in the Vale Local History Society, James might have had trouble retaining his property because costs associated with enclosure lands were sometimes too much for a farmer with small holdings.

Problems holding the land might have been part of the story behind his sons' emigration. Another reason might have been that James embarrassed and angered his children, and that cemented an urge to emigrate that the children already possessed.

Their mother and James's wife, Ann Tarrant Puzey, died in 1824 at Eastfield Farm. In 1832 James married Sarah Keen. Family stories contend Sarah was a maid in his household. Marrying a servant might have been scandalous enough, but Sarah was in her mid-twenties while James was sixty-four.

Ann and James's son James left for New York State at age forty in the same year his father remarried. Their oldest son, John, left for Ontario three years later. The other Puzey siblings stayed around for a while, but they were unsympathetic to their father's remarriage. The elder James lived until 1833, about a year after his marriage to Sarah. When he died his young widow was pregnant with their second child, and she was left destitute. The 1841 English census shows Sarah Keen Puzey living with her daughter, Phillis, aged seven, and her sons, Philip, aged eight, Richard, aged three, and Albert, aged one, in the Farringdon Union Workhouse. She was listed there as an agricultural laborer. Her last two children used the Puzey name throughout their lifetimes, but they were not James Puzey's sons. The children of James Puzey's first marriage who were still in England didn't feel obliged to help their young stepmother and her children, two of whom were their half siblings, even though they lived nearby.

—⚭—

James and Ann's son Richard did not emigrate until 1847, but he spent more time in Berkshire as an adult than the other Grand Prairie Puzeys and might have had more farming skills. Family stories describe Richard as having attended Oxford University, but Oxford is meticulous about its colleges' rosters and has no record of him. The city of Oxford, though, was nearby and thriving in the 1820s and '30s, as it is now. Many Berkshire youths would have been drawn there to find work, and Richard might have been one of them. Richard's American granddaughter, Matilda Puzey Hinton, wrote that as a boy Richard was a shepherd. That's likely, since most rural English communities had plenty of sheep. Matilda also wrote that he was a baker and helped his brother in London. Since his brothers in London were textile merchants, it is more likely he helped his brother Joseph, whose furnaces for both bricks and bread just outside Stanford in the Vale, are still remembered.

Matilda also writes that Richard lived with Quakers in London where he acquired the use of "thee" and "thou" in his conversation. Richard, however, had ample opportunity

Getting to Grand Prairie

to learn such habits closer to home. He lived among a number of Quaker families in his native Berkshire. "Thee" and "thou," moreover, were in common use among those of all religions in rural England during the nineteenth century, as we know from Tennyson's poems.

In 1825 Richard married a Berkshire girl, Elizabeth Kirkpatrick. Richard carried her Bible with him across the Atlantic. Elizabeth's name, the date of September 20, 1798, and a little handwritten poem were inscribed inside. The poem is as follows: "Elizabeth Patrick is my name, England is my nation. Sanford [Stanford] is my Dwelling Place, Christ is my salvation." The date is perhaps her birthday, but it might not be. Elizabeth's birth record was not to be found, and census records show her as having been born about 1806. The census taker, however, might have been mistaken.

In any event, according to the 1841 census, Richard Puzey, aged forty, and Elizabeth, thirty-five, were living in Stanford in the Vale with a girl, Ann Kirkpatrick, aged twelve, presumably Elizabeth's relative. Richard's profession is listed as a draper—the same as his older London brothers. What is obvious in the census is that Richard and Elizabeth had no children. Either they were unable to have children, or those children died. In 1844 Elizabeth herself died. By 1847 Richard was serving as a postmaster in Stanford in the Vale. Without a wife and children, though, there was little to keep him in Berkshire.

Sophia Puzey Church, James and Ann's ninth child, was the fourth Puzey of that generation to immigrate to America. Like Richard she did not follow their older brothers to Ontario or New York but instead settled in the Grand Prairie. Before that she had left Berkshire for London, where two of her brothers were working.

28

A Rift Among Puzeys

The children of James and Ann Puzey who stayed in England typically did not remain in Berkshire. There is no record of their oldest child, Ann. James had immigrated to New York in 1832, and John was in Canada by 1837. Their son and daughter who went to the Grand Prairie were gone by 1850, as were Philip and Phillis, James's children with Sarah Keen.

All but one of their remaining children moved to London. Sarah married a Yorkshireman and spent the rest of her life in that city. Sons Thomas and Henry made a small name for themselves as haberdashers, hosiers, and linen drapers in the neighborhood of Bow, a center of cockney life. Later in Bow, Henry ran the White Hart, a pub his children and grandchildren continued in operation until after 1900. Their daughter Jane never married and lived with Henry or Sarah the rest of her life.

Their son Joseph stayed in Berkshire and appears to have had a successful family life in Stanford in the Vale. He prospered as a brickmaker and baker, employing his furnaces for that dual purpose. He owned and occupied a little more than three acres aptly called Brick Kiln near the village and rented other lots. Joseph Puzey's wife, Beatrice Blanche, was from nearby Latteridge in Gloucestershire.

Beatrice and Joseph were able to afford a good education for their children. Census records say their son Frederick Francis, or Francis Frederick, as he sometimes called himself, worked as an accountant, which would have required special training. Their son Henry's obituary in an Illinois newspaper says he was "given the advantages of a good education, attending schools in Audlem in Cheshire and Marlborough in Wiltshire."

Getting to Grand Prairie

As relatively prosperous as Beatrice and Joseph were, it wasn't enough to keep their children nearby. Four sons, Henry, Albert, Jonathan, and Thomas, left in the 1850s for the Grand Prairie.

Of their remaining nine children, four died early and two, Frederick (or Francis) and Jane, never married. They lived with their mother most of their lives. Daughter Ann married Charles Rymer of Chepstow on the border of England and Wales. This family connection was important later to Henry when he returned from America to find a wife.

Another son, George Edwin, kept the brick kiln going after his father Joseph's death. He prospered as a farmer too. According to the 1871 census, he held two hundred acres of land and employed eleven men and three boys as farmhands, as well as a dairy maid and a female domestic servant. He lived in Stanford in the Vale at Bow House, which he built in 1865 out of red brick and tiles. The house was well fitted out with a cheese room, a back kitchen with a bread oven, and an adjoining dairy. George Edwin's younger sister Frances and her farmer husband, John Brooks, lived next door.

Roly Puzey, a descendant of George Edwin Puzey and Dee and Tom Belton's distant cousin, still lives in the area. He and his wife, Camilla, are today's version of tenant farmers. For several years they raised grass-fed lambs without hormones or antibiotics on Hill Farm, owned by Earth Trust, a nonprofit devoted to sustainably saving the English countryside. This farm is in Little Wittenham in South Oxfordshire. It is about a forty-minute drive over narrow lanes from Stanford in the Vale.

Recently Roly and Camilla moved to another Earth Trust property, Saddlescombe Farm in West Sussex. There

30

A Rift Among Puzeys

they supervise three hundred breeding ewes. They open their farm to the public for visits and also invite members of the public and children with disabilities to help them on the farm.

—⚍—

Joseph died in 1854, but Beatrice lived until 1889. After their four children emigrated, Beatrice and the children kept in touch by letters. Two sons returned to England to find wives. The parents, therefore, would have seen Albert once more, and after Joseph died Beatrice saw Henry again. Beatrice was also reunited with Jonathan, who returned to Berkshire in the 1870s to live out his days, although he did return to America for a visit at the end of the 1880s.

As the Grand Prairie Puzeys decided to emigrate, they might not have been aware they weren't the first Puzeys to leave for America. Berkshire Puzeys had come to America prior to the nineteenth century. Nathan Pusey, a president of Harvard University in the 1960s, was descended from one of these families. Puzeys inhabited many Middle Atlantic states, and there were American-born Puzeys buying and selling land in Illinois in the mid-nineteenth century before Richard arrived.

These distant relatives, however, most likely had been forgotten by the time the Grand Prairie English came. Besides, the later Puzey emigrants from Berkshire had other connections leading them to America.

31

5

Another Land of Lincoln

Will Clipson and most of the fourteen family members who accompanied him to Illinois six years after Richard Puzey immigrated were born in pastoral Lincolnshire. This county, the second largest in England, lies on the North Sea in what is now called the East Midlands. Rolling chalk hills, known as wolds, characterize Lincolnshire's northernmost countryside. These hills descend softly into the fens—the flat, reclaimed marshland to the south. A similar but smaller marshland in America gave its name to Fenway Park, the home of the Boston Red Sox.

While Lincolnshire farmers grow vegetables for the London market, they also plant sugar beets, barley, and flax. The flax flowers that bloom in June are periwinkle blue. They provide a nice contrast with the earlier-flowering yellow rape, which is grown for the oil in its seed. In midsummer the palette can change again in some quarters with the appearance of pale lilac opium poppy blossoms, grown legally for the British pharmaceutical industry.

Will was born in the fens on a bit of land assigned to the village of Miningsby, which lies in the wolds in the part of

33

Lincolnshire called East Lindsey. It is less than a mile from where the fens begin. Miningsby is not much of a village. It's unclear if it was ever much more than it is now—an intersection in the road with a farmhouse built in the late nineteenth century.

Adjacent to the farmhouse is a pasture filled with lowing cows and a footpath that leads to a gate and an enclosure overgrown with stinging nettles and cow parsley. The red-and-black tile floor of St. Andrews, the little church demolished long ago, is still there. It is where Will and his brothers and sisters were christened.

It is a peaceful setting but deceptive. At any time the screams of low-flying jets from nearby Royal Air Force bases can disrupt the idyll. Lincolnshire's location on the North Sea put it within striking distance of German targets during World War II, and more than fifty airfields were active in the 1940s. Nine RAF installations of varying types are still in use in the county.

It is obvious why these lowlands with no natural barriers were easy targets for ancient northern European mainland tribes. The earliest inhabitants of Lincolnshire are believed to have come from around the mouths of the Meuse, Rhine, and Scheldt Rivers in what is now the Netherlands. These tribes were experienced seaside dwellers. They built their villages on high ground and threw up sea banks, on top of which they ran a road they called a "ramper." Some Lincolnshire historians claim this is the root of the word "rampart." They knew how to drain land, and their descendants built windmills. It's no wonder the southeastern part of Lincolnshire is called Holland.

The Romans arrived in Britain in 55 BC. They established the town of Lincoln on an old Iron-Age settlement on the highest point of land in the area. The town enjoyed a natural harbor in the adjacent Witham River, then known as the Lindis, formed from the Celtic "lin" (pool) and "isse," which signified an island. The river's name derives from the effects of the tide, which in Roman times ran up the river to Lincoln and isolated the present county's eastern section. The Lindis gave its name to the northern part of Lincolnshire, now called Lindsey. A third part of Lincolnshire called Kesteven lies to the southwest. The town itself became known as Lindum Colonia, meaning a major colony near water, and it eventually elided into Lincoln. There is no credible evidence that Abraham Lincoln's English forebears came from Lincoln.

The Romans were in charge of what became Lincolnshire for about four hundred years. The Saxons and Angles followed, forming the Kingdom of Mercia, which also lasted for about four hundred years. Then the Danes arrived in the ninth century and established the Danelagh. Their authority was eclipsed by the Normans, who arrived in 1066.

The Danes left their mark on Lincolnshire's language. The Danish ending for a village or town is "by" (as in Miningsby), which can also mean an abode or dwelling. A writer as late as 1880 described the Lincolnshire dialect as having Scandinavian traits, with "bairn" signifying children, "holm" used to indicate an island, and "gaard" describing an enclosed yard.

The surname Clipson comes from Old Norse. Other surnames originating along England's eastern coast do so as well. Clipson has many variants. One old Lincolnshire record of such a family name comes from Thomas Clyppsam, who was christened on February 13, 1565, at Thurlby. A village in what is now Leicestershire, adjacent to Lincolnshire on the west, carries the name of Clipsham. Clipsham Hall, known

Getting to Grand Prairie

locally for its yew hedges carved into topiary, does as well. Clipsham limestone is mined in quarries around Clipsham and was used in the houses of Parliament and buildings at Oxford University. In 1783 Will's grandfather, John, was still spelling the name as "Clipsham" when he recorded a daughter's birth.

The Danes weren't the only ones to leave their mark. In the late eleventh century, William the Conqueror, in the midst of his conquering, ordered Lincoln Cathedral to be built. Most of it was completed by the fourteenth century with repairs occurring after one or another part of it fell down. Without precise engineering it was not uncommon for parts of buildings to fall down. The steeple on the church at East Keal, where Will's sister Rebecca lived, collapsed in 1853 shortly after Will and his family left for America.

By the time the American author Nathaniel Hawthorne saw Lincoln Cathedral in 1856, it looked pretty much the way it does now. Hawthorne pronounced the town of Lincoln "delightful," and he did not hold back on his praise for the cathedral. Its site on the top of a steep hill makes it visible for miles around. "I never saw anything so stately and so soft as they [the castle keep and the towers of the minster] now appeared," he wrote in *The English Notebooks*, the memoir he kept while serving as the U.S. consul at Liverpool. The nineteenth-century art critic John Ruskin called the cathedral "the most precious piece of architecture in the British Isles."

While the cathedral is impressive, so are Lincolnshire inhabitants. They are, according to local opinion, a rough and ready bunch. The novelist Patrick O'Brian described "a fierce and independent race, the Slodgers, who lived by wild-fowling, fishing, cutting reeds, raising geese, and digging turf" on yet-undrained fenland.

36

Another Land of Lincoln

Perhaps the most influential sons and daughters of Lincolnshire were the Puritans who, led by John Winthrop, left for America in 1630. They named their new community "Boston" after the seaside Lincolnshire town and Puritan stronghold, said to have been founded by St. Botolph, they left behind. Emigration was in the Lincolnshire native's blood. Between 1630 and 1640, one-tenth of the population of Boston, England, left for New England. Apparently they were under the impression that purifying the Church of England was most effective when accomplished away from English soil.

Farms dominated nineteenth-century Lincolnshire, as they do today. According to exhibits in the Museum of Lincolnshire Life, farmers of that period still used wagons decorated in an orange paint known as Farmers Glory, similar to those in the Netherlands. The county's common lands had been undergoing enclosure sporadically since the 1200s. Will's family lived on the fens, however, where canal building most affected farming after 1809. In that year Parliament ordered the fens to be drained. From then on canal building proceeded at a regular pace. This turned the marshes into arable land.

Hunters went after ducks, geese, curlews, and plovers—known locally as pyewipes. They must have hit their targets. In one nineteenth-century season, they sent thirty-one thousand ducks to London markets. They also dealt in feathers, presumably from the same birds.

Lincolnshire folk traditionally fished in the marshes, collected cockles and shrimps, and grew berries on higher ground. They sent their pickings to the London and Yorkshire markets, first by canal barge and later by rail. As Will and his future Lincolnshire fellow immigrants were growing up, however, they saw hunting, berrying, and other

37

historic fen activities decline. As more land was drained, the fen farmers gradually grew more potatoes as their main crop. Barley and other grains were grown in the wolds.

Not all Lincolnshire inhabitants were farmers. Alfred, Lord Tennyson, only three years younger than Will, was born and raised in Somersby. His father was a rector, and Tennyson was well educated, attending Trinity College at Cambridge. It is doubtful anyone from the Lincolnshire group destined for the Grand Prairie was intimately acquainted with him, but judging from Tennyson's writings, he definitely knew Lincolnshire farmers.

A century before, the Anglican clerics John and Charles Wesley, founders of the Methodist movement, had grown up in Epworth in northwestern Lincolnshire. At least two of the Grand Prairie English were said to have followed Methodist precepts before they emigrated. In 1643 Isaac Newton was born in Woolsthorpe-by-Colsterworth, which was about thirty miles west of Boston. A list of influential Lincolnshire folk also includes Margaret Thatcher, England's first female prime minister, born and raised in Grantham, a few miles north of Isaac Newton's birthplace.

The Lincolnshire folk who have molded literature, philosophy, politics, and science are impressive. Lincolnshire, however, is out of the mainstream of England. Although it is pretty, it is not a popular tourist destination. With its big sky and gently rolling, cultivated fields, it lacks drama. It is more like the Grand Prairie than its emigrants might have initially realized—peaceful, flat, quiet, and enduring.

6

Lincolnshire Ties

William Henry Clipson left for Illinois in a hurry in 1853. He managed to gather fourteen family members to go with him, including his second wife, Matilda, their five children, and his two older girls from his first marriage.

Accompanying them were Matilda's younger sister Emma, her husband, William Dickinson, and their four children. By the late 1850s, three young men, John Carby and Richard and John Todd, had followed. William Dickinson had lived with the Todd family in the 1840s. In 1881 another Dickinson family joined them in the Grand Prairie. Except for Will's London-born children, they all came from Lincolnshire.

Compared to other Grand Prairie immigrants, Will's life is easy to follow because he was associated with institutions that kept records, and he made it into the newspapers because of businesses he ran.

Will was born on May 16, 1806, in the Miningsby West Fen Allotment, where John and Charlotte Bowring Clipson lived for forty years. He was the oldest of the eight children whose births his parents recorded in the parish register. John had been born in Market Rasen, a town north of Lincoln.

39

Charlotte came from Skegness on the Lincolnshire coast. Will's mother and father were not particularly creative in naming him. William vied with John as the most common Christian name at that time in England.

Unlike James Puzey, John Clipson was apparently no landowner. He was instead a farmworker. Such farm families in the nineteenth century typically lived in small two-room houses with mud floors and chickens in the rafters. The kitchen was often in a small building of its own. As late as the early 1900s, English laborers had to boil their food since few cottages had ovens, according to historian Ronald Blythe in *Akenfield: Portrait of an English Village.* The buildings, however, were sturdy, framed in oak, and filled in with hazel wood branches covered with mud, a technique known as wattle and daub. This kind of half-timbered construction was used also in barns and outbuildings until the mid-nineteenth century. Few of these old-style buildings survive in the Lincolnshire countryside, but in the city of Lincoln some fine examples remain.

Will and his future emigrant friends were born in an unsettling time. Six months before Will's birth, the mortally wounded Lord Nelson had defeated the French and Spanish navies at Trafalgar. This victory secured Britain's position as the most potent naval force in the world. The American and French Revolutions, however, had made men and women on both sides of the Atlantic question their notions of order, traditions, and authority. America was going it alone without Britain's protection. Britain was facing the future without America's resources. France had eliminated its monarchy in a bloody revolution, and now was ruled by a diminutive man with grand dreams of conquest. English men and women wondered if such a thing could happen in their country.

Lincolnshire Ties

The early nineteenth century, however, was a creative time too. In 1806 Ludwig van Beethoven was composing Symphony No. 4 in B flat major, which he finished in the fall. Jane Austen was writing novels. Her first, *Sense and Sensibility*, would be published in 1811. Thomas Jefferson, the third president of the new United States, was anticipating the return of Lewis and Clark, whom he had sent on their historic mission to explore the western territory purchased from France in 1803. The addled King George III was busy building his version of Kew Palace.

The Clipsons' wealthy neighbor, the naturalist Sir Joseph Banks, dropped in from time to time at his large country estate in Revesby, adjacent to Miningsby at the edge of the fens. At the estate's stone manor house and farm buildings, now in ruins, he oversaw experiments in breeding contraband merino sheep from France and Spain with England's native Southdowns, Norfolks, and Lincoln's own longwool breed. Improving England's wool was critical. By the end of the eighteenth century, one-quarter of the country's exports depended on wool manufacturing, and Spain, with its fine merino flocks, was a serious competitor.

Banks was the unofficial first director of the Royal Botanic Gardens at Kew. He was also a world traveler. He accompanied Captain James Cook, then only a lieutenant, on Cook's first voyage to the South Pacific. Captain Cook features prominently in the Clipson family story, which asserts that Will's first wife, Jane, was a descendant of that adventurer. Cook's children, however, died before producing children of their own, and there is no evidence that Will's first wife was linked in any way with Captain Cook's sister, whose descendants did survive. Perhaps Will told stories to his children about Sir Joseph Banks and his relationship with Captain Cook. Then things got confused. By the

41

time the story was written down, Captain Cook had become associated with Will's first wife instead of Sir Joseph.

As perplexing as family stories can be, there is no doubt about the stature of Joseph Banks. He was one of the first scientists to theorize about climate change. In the early nineteenth century, European summers grew colder and arrested crop development. Banks believed the breakup of a massive ice shelf on the eastern coast of Greenland caused the declining temperatures. The breakup sent icebergs south, "imparting a very sensible degree of cold to the atmosphere of the countries of Europe during the summer months," according to a letter written by one of Banks's young devotees.

Banks was also an enlightened landowner. A 1987 biography by the adventure writer Patrick O'Brian described Sir Joseph: "At a considerable loss to himself, he ran his estate on the basis of small farms, so that as many men as possible should be on their own holdings."

Such an arrangement would have given local agricultural workers a leg up in achieving some degree of financial independence. The greater stability of those workers who had a reliable tenancy would have underscored how important it was to own land or control it. While Will Clipson had no such aspirations, he would have understood such intentions in his friends.

Will's father, John, had a degree of stability, since he held a piece of land in tenancy, according to an 1838 deed in the Lincolnshire archives that describes the land around parish property in the West Fen. The deed does not, however, mention who owned the land. Given the proximity of Miningsby and its West Fen Allotment to the manor house at Revesby, that tenancy was either near or on Banks's estate. If John Clipson farmed on Banks's

Lincolnshire Ties

land, he would have benefitted from Banks's generosity. At least he would have known men who did. Further evidence about John's financial stability is contradictory. Will identified his sixty-year-old father as a gentleman on his second marriage license in 1840, which would have meant John had independent income. Was this an attempt on Will's part to enhance his own status? The 1841 census identified John Clipson as an agricultural laborer, the lowest of the English farming classes.

It is conceivable, however, that because of Joseph Banks's practices, John might have accumulated enough money to live independently. The Clipson family stories leave the impression that John Clipson was a man of some means. Except for his proximity to Sir Joseph Banks, however, there is little evidence for that theory.

John and Charlotte Clipson registered their children's births, an indication they had some disposable income, since registration in many parishes carried a fee. Only four of John and Charlotte's eight children lived into adulthood. That was not unusual, but it must have caused vigilant parents untold fear and despair. One historian estimates that in the nineteenth century, 62 percent of agricultural laborers' children died before they were five years old.

The Clipson's second child, Sarah, was born in 1808 and died at the age of two. Rebecca was born in 1811 and survived into her forties, long enough to enjoy a late marriage but no children and long enough to witness her only surviving brother leave permanently for America. Bowring, given his mother's maiden name, was born in 1813 but died shortly after birth. So did John Bowring, who was born in 1822. Charlotta, born in 1817, can't be accounted for after her christening, but Lucy Bowring, born in 1826, married George Bogg and lived until the late nineteenth century.

43

Their descendants were still living in Lincolnshire in the early 1900s. A second Sarah, born in 1819, appears in the 1841 census living with Will and his family in London. After that she can't be found either.

Stories about Will's sisters provided the seed for a discussion among Will's Illinois grandchildren. The Clipson descendants considered sending a family member to England to try to claim an inheritance they believed one of Will's sisters had left. They eventually decided against it. Such a trip would have been a waste of time and money. As the American consul in Liverpool, Nathaniel Hawthorne had to deal with many Americans who landed at his office seeking help in finding an imagined fortune. He was contemptuous of such efforts. "Besides his desire to see the Queen, [the American] has a fantasy that he is one of the legal heirs of a great English inheritance," he wrote.

—⚉—

While John and Charlotte Clipson were raising and losing their children in Miningsby's West Fen Allotment, James and Anne Barker, the parents of Will's second wife, Matilda, were trying to start a family of their own a few miles south in Boston.

Boston is located on the Wash, a wide inlet of the North Sea. Except for a time spent as part of the Hanseatic League, the town never was an important seaport because of its shallow harbor. Nevertheless, in the nineteenth century, it was a thriving county center with canals, ditches, drains, and dikes, accompanied by sluices and locks, forming a network around the city. Packet boats moved farm goods and people by canals and the River Witham, which connected Boston and Lincoln. With the advent of the railroad in the mid-1800s, the canals

Lincolnshire Ties

grew less important for transport. Boaters, anglers, and birders enjoy the narrow, languid waterways today.

Matilda Clipson's father, James Barker, was born in 1773 farther north in Lincolnshire in Tealby, which was located in the wolds. By 1804 he had moved about thirty-five miles south to Boston. There he married nineteen-year-old Anne Cumberlidge, sometimes written as Cumbridge, at Boston Stump, the nickname for St. Botolph's Church, the largest parish church building in England. Anne was born in Skirbeck, now part of the town of Boston.

Although they married in 1804, the Barkers did not have a recorded birth until 1813, when their son James was born. Two years later, Will Clipson's second wife was born and baptized Anna Mathilda. (Her name in later records varies.) Harriet, Thomas, and Henry followed in 1817, 1819, and 1820. Emma was born in 1823. She and her family later accompanied the Clipsons to America.

James and Ann prospered in some ways. England's first civil census was in 1841, and it showed the Barkers were living at Mile House between Frith Bank and Witham Bank—the edges of two canals north of Boston. On his marriage license, James was listed as a laborer, but by 1841 he had risen to the title of farmer. This meant he owned land or had a stable tenancy. The more complete 1851 census listed his farm as comprising twenty acres, and his family enjoyed a house servant.

Other Grand Prairie immigrants from Lincolnshire were also growing up and learning to farm in the first quarter of the nineteenth century. William Dickinson, Emma Barker's future husband, was born to John and Hannah Dickinson in 1819 in the village of Skirbeck, where Anne Cumberlidge Barker had been born. The 1841 census showed no John or Hannah. Their son William was listed

45

as an agricultural laborer living with their neighbors, the Todd family. John and Richard Todd were youngsters at that time, but later followed Dickinson to the Grand Prairie. No public records show what happened to William Dickinson's mother and father or his siblings, if he had any.

The differences in the arrangement of American and English farmsteads attracted Nathaniel Hawthorne's attention. In America farmhouses were scattered. In England "they cluster themselves together in little hamlets and villages." The farmland itself surrounded those settlements. Hawthorne believed in the unchanging nature of the villages. Each village was "a place where everybody had grown up together, and spent whole successions of lives and died and been buried under the same sod—the same family names, same family features, repeated from generation to generation." This, however, could be said of some American towns and villages too. Hawthorne maintained Americans would weary of this.

The different birthplaces of Grand Prairie immigrants' forebears, however, show that while Hawthorne saw unchanging populations, there was considerable change. Young men and women were constantly moving to new villages in search of novelty and opportunity. In any case, remaining in unchanging English villages was about to end for the Grand Prairie English, if it ever existed.

Even if the Clipsons benefitted from Sir Joseph Banks's enlightened ideas, young Will must have realized that if he were to farm without owning land he would always be at the mercy of the landowner. In any case, in his early life he showed no interest in becoming a farmer. Will was a restless

Lincolnshire Ties

sort, as his many later moves and attempts at different jobs indicated. Later incidents showed he took risks too.

Will must have become aware that some Lincolnshire farmers were leaving for better opportunities abroad. Between 1822 and 1824, Thomas, Kelsey, and Joseph Crackles, "a fine specimen of English farm laborers, well skilled in every description of farm labor, and particularly in the draining of land," left for Edwards County in southern Illinois according to Charles Boewe's *Prairie Albion*. There they joined a growing population of English-born settlers. They soon got farms of their own and prospered. These young men and others like them were leaving Lincolnshire and other English counties for America, Australia, and Canada, seeking adventure, status, and possible wealth. As the century wore on, the numbers of those leaving rural England would only increase.

Sir Joseph Banks died in 1820. His will left everything to his wife, Dorothea. Because they had no children, his estate was to be divided when she died between two distant cousins. It is unclear if the heirs were enlightened landlords as Sir Joseph had been.

What is clear is that in 1824, when Will was eighteen years old, he enlisted in the Royal Marine Artillery. He left for Plymouth, a seaside town in Devon, and returned to live in Lincolnshire only for a few months before he left for America.

Clipson family stories claim that Will enlisted as a soldier in the British Army and served as a messman to King William. At least one descendant maintained Will was a taster for the king. Family stories tend to exaggerate the importance of an ancestor. Nevertheless they sometimes contain a kernel of truth that might explain how the story got started. In this case the kernel lies in the location. Outside

47

Getting to Grand Prairie

Plymouth was the Royal William Victualing Yard. This was associated with Devonport, which had been a Royal Navy facility since the late 1600s. The yard provided food for the ships that sailed in to stock up. Now known as Her Majesty's Naval Base Devonport, it is still in operation and is the largest naval base in Western Europe.

The yard was under construction when Will was in Devonport. It is a stretch of the imagination to think he would have served the king. Nevertheless, if he had helped with provisioning, it would have been good training for running a pub, which he eventually did.

The Blitz destroyed most of the archives of the Royal Marine Artillery from this time, so there are few records in the British National Archives at Kew. It's difficult then to know much of what Will did in Plymouth or how such a story got started. The reliable information comes from his marriage license, which lists his occupation as a gunner. Will was married in the parish of Stoke Damerel, where the naval base is located, to nineteen-year-old Jane Shaw on June 17, 1828, when he was twenty-two. Jane was the daughter of William Shaw, a baker. Her mother was born Eugenia Elworthy.

By 1830 Will had transferred to Portsmouth, where there was also a provisioning yard. His son, William John, named after both grandfathers as well as Will himself, was born at the Portsea Naval Base in October. William John, however, was not alive when the 1841 census took place. Will's biography in the 1889 *Portrait and Biographical Album, Vermilion County, Illinois* reports he and Jane had eight children, but only two survived.

In the early 1830s, Will decided a lifetime spent in the Royal Marine Artillery was not for him. He and Jane joined countless others in moving to London, where they hoped to find opportunity and adventure.

48

7

Lighting London

The men and women who immigrated to the Grand Prairie in the mid-nineteenth century were able to finance their moves and their subsequent acquisition of land because they prospered in London during that city's explosive growth. But it took most of them several decades to accumulate that wealth.

In 1801 London was already a large city with a population of about 959,000. Acknowledging its attraction as well as abhorring it, the poet William Blake called the city "a human awful wonder of God." By 1815 London was the center of the world's economy. Its advantages and disadvantages would intensify as the population grew by 1831 to 1,655,000, an increase of 73 percent. The population of the entire state of Illinois at that time was less than 10 percent of London's.

Medieval practices that once satisfied the needs of a small city left a growing London with no comprehensive government. London was divided into parishes—neighborhoods surrounding an Anglican church. Parish vestries, the committee of men who had assumed leadership roles, were historically in charge of city services. The practice of

49

dividing London into parishes was so entrenched that when the government conducted the 1841 census, which was the first to include residents' names, it used parish boundaries in counting the population. The vestries imposed "rates" upon those living within parish boundaries to pay for supporting the poor, maintaining order, and sometimes for sewer and water lines, if a neighborhood were lucky enough to have such amenities. Private companies and Parliament increasingly supplemented or substituted services that had once been the vestries' domain, but coordination was irregular at best. This messy state of affairs made it hard to keep up with the growth.

Growth, however, was relentless, and all kinds of services, activities, and institutions arose to accommodate that. Historian Jerry White's *London in the 19th Century* put it in context: "The year 1825 will ever be memorable in the annals of London; for within that period more novel improvements, changes, and events have occurred in the metropolis than during any other corresponding extent of time." The same could have been said of several years between 1815 and 1830.

During that time the intemperate and unpopular George IV presided over Britain. He had served as prince regent when his mentally ill father, George III, was indisposed. During that period and later during his own reign, which began in 1820 on the death of his father, George annoyed members of Parliament and the populace. He illegally married a Catholic wife yet succeeded in denying more freedom to Catholic followers in England. He plunged himself into debt, and Parliament had to bail him out. He drank prodigiously and had numerous affairs. He legally married a wife his family chose. Then he rejected

Lighting London

her and fought publicly with her for more than thirty years. His subjects considered him selfish, lazy, and embarrassing. But he had a taste for architecture. King George IV and his architect and city planner, John Nash, transformed London into a modern city. They laid out new streets and squares, built beautiful buildings, created parks, and designed stylish neighborhoods for the well-to-do. "What they achieved together between 1811 and 1830 was unprecedented in its day and is unmatched since," declared White, writing in the twenty-first century.

With Nash's help, George IV converted an old mansion, Buckingham House, at the western edge of London, into an elaborate, permanent home—a palace, actually. This drove real estate speculators to move into the area with new, fashionable developments. In 1827 John Nash designed the Marble Arch, made of Carrara marble and based on the Arch of Constantine in Rome, to serve as the entrance to Buckingham Palace. It was later moved, but its grandeur and classical references expressed George IV and Nash's confidence in Britain's preeminence.

Lesser folk were also hard at work on London's modernization. Roads, sewers, and hospitals were built at a rapid rate, as were prisons and theaters. In the 1820s the Southwark Water Company and the Vauxhall Water Company were piping water from the Thames River to private households. By 1833 Vauxhall alone supplied water to more than twelve thousand houses.

During the 1820s, additional bridges built over the Thames made it easier for people to cross the river. Pedestrians and horses and carts were usually charged a toll. The gradual ease by which people could cross the Thames stimulated house building on the river's marshy south side

51

where many Grand Prairie immigrants could be found in the 1840s.

A new London Bridge replaced the old over a period of eight years beginning in 1823. The National Gallery opened in 1824. St. Katharine's Dock was built in the Thames between 1824 and 1828, increasing London's ability to handle trade. The Metropolitan Roads Board swung into action in the 1820s with new roads, some paved, that encouraged the use of horse-drawn omnibuses carrying a dozen people inside and the rest on top. Londoners bragged their city had more paved roads than Paris.

London Zoo, with the world's largest animal collection, opened in Regent's Park in 1828. Although the tradition of gentlemen's clubs had begun a century before, the 1820s saw a rapid growth on Pall Mall and St. James Street of these edifices of gambling, intrigue, and isolation from the teeming, putrid world outside.

The outside world certainly reeked. Despite its advances, London was a city of Dickensian filth and squalor. Carriage traffic in the city was horrendous, horse manure filled the streets, sewers were inadequate, and bathing facilities were scarce. Unsurprisingly, many Londoners suffered from poor health.

Progress for some meant displacement for others. To provide access to St. Katharine's Dock, more than twelve hundred houses were demolished. More than eleven thousand people were left homeless. Living conditions, even for those who were housed, were not comfortable. Most working-class families lived in one or two rooms. Charles Dickens's *David Copperfield* was set around this time. It mentions a water butt, or barrel for collecting rainwater, as a feature in the Micawber family's back yard. Given the coal fires

Lighting London

and gas-burning lamps fouling London's air, the water such families collected must have been filthy.

—⁂—

This was the London into which Henry Jones was born, and this is where he prospered. For about thirty years, he lit London—its shop windows, theaters, parks, and probably some of its streets too—using his skills as a gas fitter.

Henry was born on May 25, 1804, and christened at age five with his older brother James at Clerkenwell—a neighborhood in central London just north of the Thames. Of all the Grand Prairie English, he was the most imposing. He was a natural leader, an optimist, and an entertaining, if possibly frustrating, friend and family member.

The family story says his father, Richard, was an "extensive timber merchant" from Wales. The name Jones makes it hard to find credible records. Like the baptismal name John from which it derives, it is one of the most common surnames in the English language.

With a reported weight of three hundred pounds at some point in his life, Henry had an unquenchable appetite for living. A skilled worker, a successful businessman, a confident traveler, and a man with money in his pocket, he would land in New York in June 1849, two years after Richard Puzey. Shortly after he arrived, he would post a lively letter detailing his enjoyment of the trip across the Atlantic. He would travel to Illinois immediately thereafter, intending to return to New York two months later to meet the ship carrying twenty-five family members and friends. Those immigrants weren't the only ones who followed his lead. He was also responsible for determining the

53

Getting to Grand Prairie

destination of the Clipson and Dickinson families when
they left England four years later. His enthusiastic embrace
of emigration must have boosted the confidence of his
friends and neighbors who were considering taking such a
dramatic step.

During his teenage years, Henry learned the demand-
ing and in-demand craft of gas fitting. Sometimes he was
described as a brass founder and fitter. At age nineteen he
married Sarah Hough, also nineteen, at St. Dunstan's in the
London neighborhood of Stepney on New Year's Day in
1824. Sarah was from Derbyshire in the middle of England.
Although her home was in a rural area, her family members were not farmers. St. Werburgh, where her mother and
father, Ann and William Hough, christened her, lay in the
midst of a growing textile manufacturing industry, and the
men in her family were craftsmen.

By the 1820s, she had made it to London. She followed
her brother William, a carpenter with a specialty as a joiner,
and her brother Samuel, a goldbeater and whitesmith, a
person who works with tin and pewter. Samuel had lodg-
ings and a workshop on Fetter Lane and Brownlow Mews
near Hatton Garden, which is still the center of London's
jewelry and diamond trade. He conducted business there
throughout the 1820s and 1830s.

Sarah and Henry set about growing a family and
building a business. The births of their first four children,
all baptized at St. Anne's in Soho where they were living at
the time, are meticulously detailed in the family story and
corroborated by official records. Their son Richard was
born on December 5, 1824, at 9:20 a.m. Sarah Elizabeth was
born at 12:25 in the afternoon on October 22, 1826. Eliza
was born on February 3, 1829, at 12:25 p.m. She later kept
a charming diary revealing the relationships the Grand

54

Prairie immigrants had in London. Rebekah was born at 1:30 p.m. on November 7, 1831.

The Joneses were an active, well-educated family, as evidenced by the writings of Henry and Eliza. Four of Henry and Sarah's children died early, but they were one of the lucky couples. Seven children made it through the dangerous ages of childhood, successfully weathering accidents, influenza, cholera, and the other infections that wiped out many London families.

In the autumn of 1831, for example, cholera claimed the lives of five thousand Londoners. By 1833 twenty thousand people in England and Wales had died. In London the death rate was 25.2 per one thousand people, compared with 22.5 per thousand for England as a whole. In 1833 London was hit again, this time with an influenza epidemic. Little Rebekah Jones, who died in that year, might have succumbed to that disease. Three other Jones children—Henry William, another Henry, and Frederick—were born and died between 1831 and 1844.

Epidemics, massive as they were, barely made a dent in London's growth. Henry's business was expanding. The new roads, shops, gentlemen's clubs, and pleasure gardens needed what Henry Jones so ably provided—gas lights. It turned out Henry had entered the brass foundry and gas fitting industry at an opportune time.

In 1807 a German named Friedrich Albrect Winzer had erected thirteen gas lamps on Pall Mall. A few years later he founded the Gas Light and Coke Company and methodically started lighting London neighborhoods north of the Thames. In 1821 just as Henry Jones was establishing his business, pub owners and shopkeepers began installing plate glass in windows and illuminating them with gas jets. Goods in the windows were then visible at night, dazzling

Getting to Grand Prairie

Londoners accustomed to the dark. Henry Jones made it his business to satisfy the city's appetite for light.

His name sometimes appears in insurance records with his brother James and sometimes with Richard Jones, both of whom, like him, were brass founders and fitters and were in business at 23 and 73 Grafton Street, where Henry and Sarah lived in the 1820s. Richard's relationship to Henry cannot be verified, but it is likely, given that Henry's father's name was Richard, this Richard was also a brother. For a time Henry tried working with other partners to make his business grow faster. He and a man named William Talby were partners by 1833, and in 1836 they took on Robert Perry as a third partner.

Later in the 1830s Henry appeared in Robson's London Directory as Jones & Co., Brass Manufacturers and Gas Fitters, 6 Rose Street, Covent Garden. He listed this address as his residence when some of his children were born, so his family must have lived here periodically, perhaps above the shop. His business occupied the space for almost fifteen years, and his friend Will Clipson took it over after Henry emigrated. A writer in the 1880s had little good to say about Rose Street. Saying it was filled with "low gambling-houses," he praised city officials for pulling down old, dilapidated tenements in 1859 and replacing them with the broad thoroughfare of Garrick Street.

This was then and still is a busy part of town. It is near theaters and the produce market made famous in America in Lerner and Loewe's *My Fair Lady*. The market has moved, but other shops have taken its place, and the area remains a popular attraction. Part of old Rose Street is still there, complete with the Lamb and Flag—an ancient pub Henry would have known. At one time it was called the Bucket of Blood

56

Lighting London

because of the illegal prizefights held there beginning in the 1600s in its rather hidden location.

Rose Street now runs crookedly between Garrick and Longacre Streets near King Street. Sometimes Henry's address was noted as King Street. Near Longacre is currently the side wall of Stanford's, which bills itself as the world's largest map and travel bookshop. It helpfully lists its longitude as 0°07'28" W and its latitude as 51°30'42" N.

—∿∿—

Henry Jones's gas fitting business was thriving. He was well established on Rose King Street, as his family typically called it. He had given up his partnership with Robert Perry and William Talby in 1841, but soon his son Richard was working with him.

In 1843, when Richard was eighteen, he and Henry caught a thief. They told the story at the trial, and their words were recorded in the Old Bailey's records. George Brooks and George Mays, both nineteen, stole a silver teapot from Henry's shop near Covent Garden.

"About half-past six o'clock, on the 26th of Jan., I was standing at the door, and saw Brooks pass—I went into the counting-house, and heard a cry of 'Stop thief'—I went and saw the prisoners in custody, with this tea-pot—I believe it to be mine. I have missed such a one," reported Henry.

Richard continued the story. "I heard the cry of 'Stop thief'—I ran out the two persons passed me, from the staircase inside the house—I immediately followed—I lost sight of Mays and followed Brooks—I am sure he was one—when Brooks got into New-street he stumbled, and this tea-pot fell from his hand—I took it up and gave it to the policeman—I

57

believe Mays to be the other man—he resembles him in every respect—he had a frock-coat on—I did not see his face."

A passerby stopped one of the thieves, and the other fell down. After all that excitement, the thieves were lucky. They received a six-month sentence, which was lenient at a time when infractions of that degree caused many to be transported to Australia.

Henry's skill at landing good jobs was exciting in its own way. One of his most prominent tasks was illuminating the Gothic Bridge at the Royal Surrey Zoological Gardens, south of the Thames. It opened in 1846 festooned with gas jets. The family history said that Richard worked with Henry, serving as "a foreman in the factory." Henry used a drawing of the Gothic Bridge in an advertising flyer the family saved. The flyer also lists some of his other work too. Most notable are many theaters still popular today.

Henry and Sarah were not yet finished having children. Like most English couples, or American couples of that time for that matter, they married when they were about twenty years old and had a child every two years or so for up to twenty years. The Joneses' daughter Emily was born in 1836 and Louisa in April 1842. In May 1844, Sarah gave birth to the second Frederick. He was given the same name as the infant son who died in 1841. Their last child, Arthur, was born in July 1848, and he survived. Sarah was forty-four at his birth.

With so many births, what did parents of that time tell their children about the delicate subject of where babies come from? Perhaps they followed the practice Hawthorne described in his English journals. "As regards the mystery of birth, we have told Julian and Una, at the advent of Rosebud, [the Hawthornes' three children] that she was sent from Heaven, as they themselves had been. Julian said this morning that he could not remember when he came down from Heaven, but that he was glad he happened to tumble into so good a family. It is queer that he is still satisfied with this first explanation of his origin—now in his tenth year."

—᙮᙮᙮—

When Henry emigrated, his brother James stayed in London working as a gas fitter, and no other siblings or cousins emigrated with him. However, members of two families related to Sarah relocated with the Joneses to the Grand Prairie. Sarah's brother Samuel Hough left for St. Louis a couple years before Henry left, but two of his daughters, Mary Ann and Elizabeth, accompanied Sarah when she led the party of twenty-five immigrants. The girls eventually joined their father.

The Hough surname's origin has two explanations. It is either associated with a local place, as in "of the hough," or its source might be in the word "haugh," which means a hill or mound. It was a common name in London's jewelry district, so Samuel and his wife could not be pinned down in London's records. Several Houghs named Samuel were jewelers or smiths. They were about the same age with children bearing similar names. What is certain, though, from Eliza Jones's diary is that the Joneses and the Houghs kept close company while living in London. After a bit of moving back and forth between England and America, these Houghs eventually settled permanently in St. Louis. They became a magnet for a few Grand Prairie immigrants when they later moved west from Illinois.

The other future Grand Prairie immigrants related to Sarah Hough Jones were the Bentleys. Thomas Bentley married Eliza Hough in 1840 after Thomas's first wife died. Eliza was the daughter of Sarah Jones's brother William Hough.

Thomas and his younger brother James made their living as curriers—skilled workers who finished leather before it went to the shoe or saddle maker. Thomas was born in what now constitutes London. At the time of his birth in 1804, however, his community of Battersea was at the southern edge of the city. Farms growing produce for market filled the area. By 1840 their father, Thomas, was working as a customs officer.

In 1826 Thomas Bentley married Sarah Holland, called Sally. She gave birth to Sarah in 1828, Thomas in 1831, and Fanny in 1836, before dying in 1839. Sally's London family was in touch with her children in Illinois by letter at least into the 1860s. Those letters confirm London relationships between the Joneses and their relations and other Grand Prairie immigrants of the time.

It is hard to know Thomas Bentley well. He wrote no surviving letters. He can't be found in London newspapers of the day. His descendants were left with no family story. It's not even clear that all his descendants are actually his.

8

Ladies and Gentlemen

The Churches were another sizeable London family whose members chose to immigrate to the Grand Prairie with the Puzeys, Bentleys, and Joneses. Of all the Grand Prairie immigrants, they were of the highest social class. Their connection to the Puzey family was through Henry Church, who married Sophia Puzey. Proof of their connection to the other London families headed to the Grand Prairie at that time was revealed only by a close reading of family letters.

It is a stretch to think that almost one hundred people from London would arrive within a short time period at the same few acres of land in a large but lightly inhabited foreign country without having known one another before. When they all bought land within shouting distance, it is easy to suspect some part of this was planned in advance.

Yet the London connection between the Churches and Puzeys and the other Grand Prairie-bound families was not immediately obvious. The Churches lived in a different world from the Joneses and Bentleys. Most of the Churches occupied the professional class rather than that of skilled workers. They lived in a better neighborhood. Their lives

Getting to Grand Prairie

were unlikely to intersect in a sprawling metropolis. The names "Church" and "Puzey" are not mentioned in either Eliza Jones's diary or Henry Jones's letter describing his Atlantic crossing. The family stories don't mention the connections until they are living in Illinois.

But one letter does. Church family members in Illinois had saved nineteenth-century letters from relatives in England, and these shed light on the family's history, personalities, and concerns. Fanny Holland Palmer, the sister of Thomas Bentley's first wife, Sally Holland, never left England. But she corresponded with her namesake and niece, Fanny Bentley, who married Adolphus Church in Illinois. In the 1860s, Aunt Fanny wrote that she had never met Adolphus, but she had met his brother Albert, who had left England in 1850. The letter is not specific about the meeting, and it does not explain the connection. It does confirm, however, that members of the Jones-Bentley families and the Church-Puzey families had known one another prior to emigration.

The Church family, headed by Thomas Church and his wife, Elizabeth Dixon, lived in the early part of the nineteenth century in the parish of St. Mary Somerset—a maritime neighborhood on the north bank of the Thames in London. This was and is the city's financial district. Thomas was a cabinetmaker and later a sugar merchant, and Elizabeth was a daughter of William Dixon, the manciple, or the person in charge of purchasing and storing provisions, at Charterhouse, an ancient school, hospital, and chapel that presently is located outside of London proper. Elizabeth Dixon's mother, Clementina, was the institution's cook. When William died, Elizabeth Dixon Church's brother William took his father's place as the manciple.

62

Ladies and Gentlemen

Thomas paid church officials two hundred pounds for a special license so he and Elizabeth could marry without waiting several weeks for the banns announcing the upcoming marriage to be published. They married two days after they received the license. No reason was given for the special license, but it was often used when the woman was pregnant, one of the couple was not a member of the Church of England, the couple were members of different dioceses, or there was family opposition to the marriage. Such licenses, however, carried with them a certain social status since they were expensive, and members of the upper classes often bought licenses. Elizabeth and Thomas had the means to pay for such a dispensation. Several of their children also bought such marriage licenses when it came time to wed.

Thomas and Elizabeth Church's family became world travelers. Their son Thomas grew wealthy as a member of the East India Company, and he and his family spent much of their lives in Singapore. Son William left England for Oswego, New York, sometime after 1833. Another son, Charles, immigrated to Australia. Their youngest son, Henry, relocated to the Grand Prairie with his nephews and wife Sophia's Puzey relatives.

Thomas and Elizabeth's oldest son, George Zephaniah Church, born in 1796, stayed in England. But his sons went to the Grand Prairie. George was a bank clerk, a dedicated gardener, and a pigeon fancier. In one letter, he described a fish pond, five feet deep, that he had worked hard on at the bottom of his garden. He stocked it with about thirty fish of several types, but few lived through the winter.

George Zephaniah is said to have married well to Elizabeth Lydia Draper. Like his parents, he and Elizabeth watched their children head for faraway places. Their daughter Adeline married a Portuguese-born businessman

63

Getting to Grand Prairie

and moved to Brazil. Their other daughters, Emma and Alice, remained in England, but their three sons, George W. F., Albert, and Adolphus, left for the Grand Prairie. George and Elizabeth's relative wealth enabled them to help those sons acquire farmland. They sent money for equipment as well.

The boys had been sent to boarding school on the outskirts of London. George was an amateur painter, and Adolphus, who was called Dolf or Doffy by his family, was fond of poetry. The boys went separately to the Grand Prairie, but each did so between the ages of sixteen and eighteen. Few youths of that age now set out alone on a voyage halfway around the world.

—⚏—

During the 1830s and '40s, moving between London and the rest of England was easier than it had ever been. Charing Cross had been declared the official center of London in 1831, enabling officials to measure distances with greater accuracy. More than four hundred horse-drawn coaches connected London with the rest of Britain, making it possible for Londoners to visit country cousins more easily than at the beginning of the century. In 1832 the Margate Steam-Boat Company carried one hundred thousand passengers a year on three steamers that left from the newly built St. Katharine's Dock on the Thames.

The greater ease with which people and goods could travel gave shopkeepers better access to merchandise, which Henry Jones's gas lights illuminated. Sophia Puzey's brothers had moved from Berkshire to London and were taking advantage of such new developments. London's shops, with their gas-lit windows, were said to be the best

64

Ladies and Gentlemen

in the world, and Sophia's brothers, and perhaps Sophia herself, were part of that world.

Her older brother Thomas and youngest brother Henry were both merchants. Over the years, census records and trade directories show they dealt variously in dry goods, linen, wool, hosiery, and haberdashery. No records show that Sophia worked in their shops, and she was not apprenticed to her brothers in the same way their nephew Henry would report he had been. Nevertheless, it would have been natural for her to help out and learn how to run her own grocery, especially before she married.

By 1830 Henry Puzey's business, Puzey and Company, was on the High Street in the East London neighborhood of Bow—the center of cockney life. Like Henry Jones, Henry Puzey tried partnering with various people, and sometimes he ran branches of his dry goods businesses from several locations. He had formed Puzey and Co. with a man named Ebenezer Stewart, and he was a partner with his wife's relative, Frederick George Lansdown. Together they formed Lansdown and Company, which briefly shared the same Bow address as Puzey and Company. The younger Henry Puzey, a son of Joseph and Beatrice Puzey and a nephew of Sophia Puzey Church and Henry Puzey, was apprenticed to the older Henry and then worked for him after his apprenticeship was over.

~~~—

During the 1830s, London continued to be an exciting but challenging place to live. Penny Dreadfuls, as the cheap newspapers that followed lurid events were called, were entertaining Londoners. Charles Dickens was writing *The Pickwick Papers* as well as working for the *Morning Chronicle*,

Getting to Grand Prairie

one of the many legitimate newspapers in London that everyone avidly read.

The most dramatic event in the 1830s that the Grand Prairie English shared with their fellow Londoners occurred on October 16, 1834. On that date a few government employees burned wastepaper in the cellars of the Palace of Westminster. The fires got out of control and burned the House of Commons, which was lodged in Westminster, to the ground. Londoners hurried out into the night to watch the fire, which lit up the entire city. They might have seen Joseph Mallard William Turner sketching frantically as the buildings burned. The next year they could relive the conflagration by visiting the Royal Academy where one of Turner's oil paintings based on those sketches was shown for the first time.

George IV had died in 1830, and his younger brother, William IV, was crowned king. William died seven years later, making room for his niece. Victoria moved into Buckingham Palace and soon ascended the throne. That same year an influenza epidemic resulted in funerals for a thousand people on only one day. Still, thousands of the living turned out for Victoria's elaborate coronation in 1838 despite the risk of catching one of the many diseases that hung over London with the smog.

—⚉—

Sophia Puzey married Henry Church, the youngest son of Thomas and Elizabeth Dixon Church, in 1832 at St. Mary Somerset, where Henry had been baptized on October 11, 1807. The marriage introduced the connection between the Puzey and Church families that continues to this day. At a time when both men and women typically married

Ladies and Gentlemen

between the ages of eighteen and twenty-two, Sophia, thirty, and Henry, twenty-five, were old for a first marriage.

It is possible that Sophia met Henry through her London-based brothers. At the time of their marriage, he was a linen draper, the same occupation her brothers listed on several documents. Henry was less successful than other members of his family. Instead of making a fortune in Southeast Asia like his brother Thomas or leading a comfortable life as a London banker like his brother George Zephaniah, Henry eventually ran a grocery, and there are hints his wife was the force behind whatever success he had.

Henry and Sophia's first child, Henry Charles, was born in November 1832 in Tottenham, which still had remnants of rural life, since the railroad lines had not yet reached that section of metropolitan London. Ann Sophia was born two years later in Dalston, a part of the London borough of Hackney. This was where Henry's brother George Zephaniah and his wife, Elizabeth, lived. It is possible that Sophia and Henry were living with them. Alfred was born in 1836 but died a year later.

By 1837 Henry and Sophia were living at 10 Mile End Road, where they kept a grocery. Thomas was born in 1838, Jane in 1840, and James in 1842, though he lived only a year and a half. Sarah, born in 1844, survived. London trade directories list Henry as a cheesemonger at Mile End Road in the late 1840s, and they lived there until they emigrated.

Mile End Road is in Stepney. It was a gritty working-class neighborhood not far from Sophia's brothers. Henry and Sarah Jones had lived there off and on. It must have been a good place for shopkeepers and tradespeople. In 1834 Charles Henry Harrod launched a wholesale grocery business there that grew into one of the largest department stores in the world.

Getting to Grand Prairie

Even though Henry and Sophia can be traced through official documents, the family is missing a few pieces in its story. In the 1841 census, Sophia appears to be the one who was primarily responsible for the shop. She was listed as having the occupation of "w. of Church," or wife of Church. Their son Henry Charles was eight years old, and daughter Ann Sophia was six. The younger children, Thomas, three, and Jane, not yet one, were also listed. Sophia had three servants or employees living with her, and that included one "shopman" to help in the family's grocery business. Henry Church, however, was missing from the family home.

Neither was Henry with the family when they crossed the Atlantic, and it is not known when he made the trip. But it was before May 1849 because on the eighth of that month, he bought farmland in the Grand Prairie. Their son Henry Charles had died in 1844, and Henry was in Illinois when Ann Sophia died in the spring of 1849. Henry played no role in the family stories or in any of the letters the Churches or Puzeys passed down.

Sophia must have found it comforting to have her brothers nearby during the year or more that Henry was in America and she and the children were still in London. Their last child, Sarah, had been born in 1844. Sarah's birth was recorded at the Kingsland Independent Chapel, founded by a Congregationalist minister. This indicated that this Church family had become nonconformists.

68

9

Another Kind of Life

During the 1830s and 1840s Will Clipson took a different route from his eventual London friends. He was restless, and he didn't like working with his hands. Some of his choices proved riskier than those of his good friend, Henry Jones, or the heads of other Grand Prairie families.

Will and Jane moved to London in the early 1830s. At first they lived north of the Thames on Charles Street in Westminster, a healthier district than some. According to a newspaper report, Westminster had had only thirty-seven cholera deaths over an unspecified period of time. This was compared to Southwark, home to poorer residents, which had suffered 409 deaths. The same newspaper that reported the incidence of cholera deaths also noted that thirty-eight bodies had been found in the Thames.

That isn't surprising. The Thames River, the source of some of London's drinking water, was notorious for its poor condition. "He who drinks a tumbler of London water has literally in his stomach more animated beings than there are men, women, and children on the face of the globe," declared the Anglican cleric Sydney Smith.

Getting to Grand Prairie

It was not only illness that plagued London. Crime also was a problem during the city's period of swift growth. Parish vestry members were in charge of police protection and night watches, but this responsibility overwhelmed them. A few small police forces helped out, but they were restricted to their own neighborhoods. The army helped out too, but Londoners were as wary of "redcoats" as Americans had been.

With experience in the Royal Marine Artillery and a written recommendation from a lieutenant at the Plymouth naval yard, Will signed up shortly after his arrival in London to become one of Sir Robert Peel's bobbies.

Peel was a member of Parliament who would later twice become prime minister. In 1829 in his role of home secretary, he addressed lawlessness by founding the Metropolitan Police Force. By 1830 there were thirty-three hundred police officers, or Bobby's men. The name soon changed to bobbies, a tribute to Sir Robert. The Metropolitan Police Force, with its officers clad in smart blue uniforms even when they weren't on duty, became fully operational in 1831, only a year before Will joined. The kind of person who became a bobby was at least 5'7" and "not a gentleman of the retired officer class" but a person who possessed a "perfect command of temper" and a "quiet and determined manner." Will and his fellow officers were also known as "coppers" for the copper-clad truncheons they carried. Soon this nickname was shortened to "cop."

Peel established the force's headquarters next to Great Scotland Yard, where the Scottish kings historically stayed when visiting London. The location eventually gave the police force the nickname of "Scotland Yard." Even today this name contributes to the atmosphere in many popular detective stories.

70

Another Kind of Life

Peel's police force came into being in the nick of time. The fear of unruly mobs was pervasive in London. After all, French revolutionaries in July 1830 had once again toppled their monarchy, and this time it took only three days. Everyone was afraid London could become another Paris.

Yet policing was a difficult job. At first the officers were unpopular. Londoners of the lower classes were restless and quarrelsome and participated in strikes and other noisy gatherings. They were particularly edgy when dealing with authority, so the job was challenging.

A bobby's pay was a guinea a week. That was equal to one pound, one shilling or twenty-one shillings. The force also contributed an unspecified amount to a few living expenses. That pay was adequate to house and feed a family, since a small house in central London at this time could be rented weekly for seven shillings, six pence. Charles Dickens's parents paid twenty-two pounds a year in rent for a four-room house in Camden Town in central London. Will Clipson could have added to that salary by hiring himself out to private employers on an hourly basis, as present-day police officers do.

What he couldn't do, however, was vote. "It was well known that the Metropolitan Police were not to have the right of voting," according to a barrister who claimed in a trial that the City of London's police force, in contrast, did have that right. Arguments over who could vote occupied a lot of time in London in the 1830s. Those who eventually headed for the Grand Prairie would have been aware that some Londoners, because they could vote, had more say over civic matters than others.

In 1832 Parliament passed the Reform Act, which did away with some egregious practices such as the "pocket" borough, where as few as seven voters controlled by a local

71

squire would send two representatives to Parliament. It also extended voting rights to any man whose real estate was worth at least ten pounds.

Whether they were able to vote or not, Will and the other Grand Prairie English families would have found self-government familiar when they arrived in Illinois because of their experiences in London. That city was full of popular newspapers that followed Parliament's actions and local affairs. Even the most casual readers would be familiar with their local officials' duties.

In addition to expanding voting rights, London was becoming more civilized by modern standards in other ways. Public whippings and the pillory were phased out, as were public executions, although these were not officially eliminated until the 1860s. The Slavery Abolition Act of 1833 ended slavery in the British Empire.

Will and Jane Clipson finally had a child who survived. Catherine, called Carrie, was born at Charles Street in 1835. Charles Street is now a lovely street in Mayfair. Jane Clipson followed in September 1836. She was named after her mother and also survived. But Alexander, born in 1838 and named after Jane's brother, did not.

By the time of Jane's birth, the Clipsons were living on Paradise Street in Lambeth on the south side of the river. This neighborhood was undergoing the nineteenth-century version of gentrification. Slums and garden vegetable farms were turned into streets with small terraced houses occupied mostly by skilled workers. While Paradise Street still exists, 1950s and '60s apartment blocks now line it since this neighborhood was leveled during the Blitz.

Another Kind of Life

The Clipsons' stay on Paradise Street didn't last long. By Jane's christening, which occurred two months after her birth, they had moved a few streets over to Carlisle Lane. Between 1835 and 1851, they moved at least ten times. Most of the other Grand Prairie immigrants—in fact, most Londoners—displayed the same restlessness. Wanting to be closer to relatives or jobs, finding better, roomier quarters, or being displaced due to civic improvements, especially railroads throughout the expanding city, triggered their moves. For example, building the Waterloo extension, near where the Clipsons lived, required the demolition of seven hundred houses in the 1840s.

In the late 1830s, Will began serving as an inspector for the New London Gas Company, a job that lasted for about a decade. That, at least, is the story put forth by the 1889 *Portrait and Biographical Album, Vermilion County, Illinois.* This volume makes a connection between Will Clipson and Henry Jones in London that Eliza Jones's diary and Henry's shipboard letter confirm. The album reports that both Henry Jones and William Clipson worked in the gas industry in London. It is likely they met through their work.

The biography spells out some of the jobs William Clipson sampled before he turned to farming in America's Midwest. He also managed a hotel, his God-fearing Methodist descendants solemnly reported in both the biography and their family story.

Some hotel. In the early 1840s, Will Clipson began running pubs. As a police officer, he would have been friendly with many pub owners, since altercations frequently took place in those establishments. He could have learned from them how to manage such a business.

In July 1839, Jane and Will Clipson were living on Devonshire Street in Lambeth with their daughters when

Getting to Grand Prairie

Jane died of "apoplexy," a term usually used for a stroke. Since she was only thirty-one, it is more likely that she died of a hemorrhage, possibly brought on by a miscarriage or another birth. On her death certificate Will reported his occupation as gas inspector.

Jane's death left Will with two little girls, aged four and almost three, to care for. It took little time for Matilda Barker to enter the picture. The family story says Matilda was a maid to either Queen Victoria or Will's wife Jane. It's probably safe to assume that Matilda's job, if true, was with the latter. Men commonly married household maids when wives died, as evidenced by Sarah Keen and James Puzey's May-December marriage.

Matilda, however, just as easily could have been summoned from Lincolnshire after Jane's death to help an old friend or a friend of a friend. Clipsons and Barkers lived within a few miles of one another on the fens. Even if they didn't know one another directly, it was probably easy for Matilda to hear of a job available taking care of a household with two motherless girls. Many rural young women took servants' jobs in London since they saw London as a place for opportunity and adventure as much as young men did.

It isn't known if Matilda found adventure. What she did find was that she was pregnant.

Will and Matilda didn't marry until November 5, 1840, at St. Mary Newington. She was already seven months along. Were they waiting to see if she miscarried, and then they wouldn't feel obliged to marry?

Their marriage certificate shows Will was working as an ironmonger—close enough to the gas inspector he seemed to have been—and living at 167 Lambeth Walk. Matilda Ann, as she called herself on this document, was nearly twenty-six. She did not list an occupation but reported

74

Another Kind of Life

herself living in Clapham Road, just south of Will's address. Will was ten years her senior. Harriet Barker served as her sister's witness. This is the certificate on which Will identified his father as a "gentleman."

William Henry Clipson Jr., or Billie, as he was called, was born in January 1841. The family story maintains that upon Billie's birth, Will gave Matilda a gift of "a candlestick holder that had on it a snuffer, controlled by a slide, just new in household appurtenances then."

Did Matilda appreciate the candlestick holder? Would she have preferred a piece of jewelry or a fashionable item of clothing instead? Was it Will's idea to put off the wedding in case Matilda miscarried? Or was it Matilda who was hesitating?

One story of a gift is curious. One of Billie's descendants claimed that Prince Albert and Queen Victoria gave the family a silver cup when Billie was born. An older Clipson family member now living in Texas recently said there was a silver cup, but it has been lost "because some of the relatives are worthless."

While the story of royalty handing over a silver cup to Will Clipson's son is hard to believe, it is also provocative. How do such stories get started in families? It makes one wish that every person kept a diary to be handed down to future generations—not that diaries are accurate or complete either.

The 1841 census showed that Will and Matilda had moved a few streets over to Belmont Place. Matilda was recorded as Matty. Will's occupation was listed as gas inspector. Will's girls by his first marriage, Carrie and Jane, were living with them. So was Billie, Matilda and Will's five-month-old son. Will's sister Sarah, twenty-one years of age, was also in residence. Perhaps she was helping with the children.

75

Will's father and mother, John and Charlotte, were listed in the same year in the Miningsby West Fen Allotment's census. Will's sister Rebecca, aged thirty, was still living with their parents. The census taker was less impressed with John's status than Will seemed to have been and recorded him as an agricultural laborer rather than a gentleman.

Will and Matilda Clipson were luckier in their children's survival than Will and his first wife had been. Billie lived, but at some point he suffered from rickets and had to be sent to Matilda's parents' home in seaside Boston to recover. Rickets is a deficiency of vitamin D that causes bones to malform. It was a common problem among children who lived under London's soupy skies. The Clipson family story tells of Billie's enjoyment of the sea and the happy occasion when his grandparents' St. Bernard rescued him from drowning. Perhaps that story or something like it is true.

The Clipsons' second son, John, was born in 1842 but died shortly thereafter. Their third son, John Clarence, called Jack, was born at Bird-in-Hand, Garden Row, in Southwark in April 1843. This is the first association of a pub with the Clipsons. Southwark, where the Joneses, the Bentleys, and the smaller Grand Prairie families also lived in 1841, was described fifteen years later by Hawthorne as "a course, dingy disagreeable suburb with shops apparently for county produce, for old clothes, second-hand furniture, for ironware, and other things bulky and inelegant."

While the Clipson family story mentions that Will first had an establishment called the Skag's Nest, there is no record of such a pub in London around that time, and London pubs are well documented. The name makes sense for Will, though. Skegness was where his mother came from, and the storyteller might have confused Will's mother's hometown with a pub name.

Another Kind of Life

Running the Bird-in-Hand seemed the job for which Will was cut out. As a former soldier and police officer, he would have been well equipped to handle drunken behavior, which the *Morning Chronicle* reported at the Bird-in-Hand several times before Will took it over. If he actually had had something to do with provisioning while stationed at Plymouth and Portsmouth, he might have been versed in how to stock a pub kitchen for serving meals. Billie Clipson reportedly told his American-born children that Prince Albert came many times to the pub, and that his father and the prince were great friends. Maybe, but there is no evidence to support it.

Pubs, short for "public houses," offered food, beverages, music, dancing, and even theater. Some had rooms for let. They were among the liveliest places in London, especially before the development of professional music halls. They were a center of community life. The patrons came from the neighborhood, and pubs, with their commodious spaces, were often the settings for trials and vestry meetings, which took up civil as well as religious matters. By 1848 there were about eleven thousand pubs in London. Historian Jerry White had great praise for them: "Of all the institutions of civil society in nineteenth-century London, the pub was without question the most versatile and innovative," he wrote.

Henry Jones's shipboard letter expressed regret at leaving Will and their activities, which no doubt involved pub life and perhaps skittles—a popular pub game. Will, however, introduced another form of entertainment that filled his pockets but eventually caused him trouble. Will ran betting games.

The family stories about how long Will served as a gas inspector aren't consistent with official documents, but it is

77

possible that his day job as a gas inspector continued while he ran the pub. He might have been saving money, but it was not for emigration. He was enjoying himself too much. The British Library's newspaper collection confirms this. Most of the Grand Prairie English did not make the newspapers, but Will Clipson did.

For example, on October 22, 1843, he placed a classified ad in *The Era*.

TO THE SPORTING MILLION.—SOMETHING NEW, at. W. Clipson's, the Bird-in-Hand, Garden-Row, London-road, Derby Sweeps for 1844. 25s. 6p. Sweep, three horses each, prizes £70, 35, 25, 12, 10 and 6 and £16 to be divided. 13s. Sweep, three horses each, prizes £40, 15, 10, 8, 5 and 33 and £8 to be divided. 6s. 6d. Sweep, two horses each prizes £20, 10, 4 and 3 and £5 to be divided. Rules to be had at the Bar. Prizes free from all deductions. Post office orders attended to. To be Drawn when full.

Such advertisements as this make it easy to follow Will Clipson as he changed pub locations throughout the 1840s and early 1850s. Will had a good thing going, and for at least a decade, he was not inclined to abandon it.

This advertisement is typical of those he placed throughout the years. Derby Day now takes place on the first weekend in June at Epsom Downs in Surrey, just south of London. It originally coincided with Whitsunday, or Pentecost, seven weeks after Easter. This day celebrates the Holy Spirit descending upon the apostles after Christ's resurrection. As a day of great jubilation, it was perfect for a horse race in which one might come away with winnings.

Another Kind of Life

Will was only one of many pub owners offering sweep-stakes for the Derby and other races. The Duchess of Kent pub near Dover Road was offering dinner and wine in celebration of the sweeps later in the year.

Will's family thrived with the arrival of James in July 1844. The Clipsons were living at Walnut Tree Walk in Lambeth, around the corner from the Bird-in-Hand. On this christening record Matilda supplied her name as Ann Matilda. Will gave his occupation as "licensed victualler." Hawthorne described the kind of meal Will Clipson might have served at his pubs. In 1855, while at a London pub, Hawthorne ate bacon, greens, a chop, and a gooseberry pudding washed down with a glass of ale.

In August 1844, a classified ad in *The Era* revealed that Mr. J. P. Hutchings was now running the Bird-in-Hand. But it didn't take long for Will to appear at a different location managing bigger winnings. In early December 1844, Will was at the Globe Tavern on Dover Road off Borough Road, just north of what is now the quaintly named Elephant and Castle tube stop. Dover Road runs, of course, to Dover on the coast, and Dickens put David Copperfield on this "long, dusty track" as he ran away to his aunt's house.

The area now has little to recommend it. Heavily bombed during World War II, Elephant and Castle and Borough Road have been disastrously rebuilt with busy highways, grim postwar architecture and unfriendly underground pedestrian passages.

A few vestiges of the past are still around, though. The Trinity, a "gastropub," looks as if it came out of Tudor England. St. George the Martyr, where several children of the Grand Prairie immigrants were christened, remains an Anglican church, but it devotes its sanctuary on Sunday

79

nights to the Christ the Resurrection Church, patronized by lively African immigrants who worship in white robes and red sashes and hats, accompanied by throbbing music.

From Borough Road Sir Norman Foster's widely acclaimed curved-top office building known as "the Gherkin" is visible. It actually looks more like a torpedo than a pickle, but perhaps a weapon was too close to home for a people who had endured the Blitz. No Globe Tavern can be found, but there is a Dover Castle on present-day Great Dover Street.

The Dover Castle was the name of the pub Will was running in 1847, according to a report on October 11 of that year in the *Daily News*. At a trial in Lambeth, G. Smith, a linen draper's assistant, was charged with selling a Mr. J. Park a forged ticket for a racing sweepstakes. Mr. Smith claimed he had bought the ticket from a Mr. Brown and had no idea it was a forgery. Unfortunately, the police had found on him a number of tickets similar to the one he sold to Mr. Park. Will Clipson, identified as the proprietor of the Dover Castle on Dover Road, testified that three of these forged tickets had been presented to him for payment. Will was quoted as saying that "frauds to a very considerable extent had been and were continued to be practiced on the public by means of forged tickets." The judge refused bail.

10

Gathering for the Grand Prairie

By the late 1840s, the adult members of the group that ended up in the Grand Prairie were mostly well settled in London. They had built businesses and established families. They could imagine their futures in London. They would work at the same tasks for the rest of their lives. Their children would have about the same advantages and disadvantages they had had. Their homes were a little more comfortable than some, but London itself was foul, gritty, full of disease, and likely to remain so for the foreseeable future.

There would also have been other disadvantages. Class profoundly separated the English at that time, and some say it still does today. Upper, middle, and lower classes had different accents. They took different forms of transportation, lived in separate neighborhoods, and shopped in different shops. They dressed differently, entertained themselves in different ways, and attended different schools.

While one might argue that today's class divisions in America fit the same description, there is an important difference. In both present-day and nineteenth-century America, one can and could move between classes with inclination, education, luck, and sufficient funds. That was not true in England in the 1840s.

Yet it would be a mistake to believe these conditions unduly frustrated the Grand Prairie immigrants. They were accustomed to such a system. They lived in the working-class neighborhoods. Two stories give clues to their manner of speech. Virginia Wallen, an older resident of the village now known as Catlin and a descendant of the English Bentleys, said her father, born in 1892, told her stories of their neighbor, the adult James Clipson, Will Clipson's third surviving son.

Mr. Clipson, she said, amused her father as a child when their neighbor Mr. Ogg drove by in his horse and buggy. "Good morning, Mr. Hogg," Mr. Clipson is reported to have said. "How are your 'ogs?"

A chronicler of a party Henry Jones gave in the 1850s in America mimicked Jones by saying he was "'appy."

Dropping "haitches" and supplying them in words where they didn't exist was the uncultured habit Professor Higgins tried to eradicate in Eliza Doolittle in George Bernard Shaw's *Pygmalion.* Later this became familiar to Americans as *My Fair Lady.*

James Clipson was pulling Mr. Ogg's leg and entertaining young Bentley with an accent closer to the lower classes than the upper. Given the neighborhoods in which he grew up, it is likely the accent was his own.

Social class constraints might have played a role in the decision of the Grand Prairie English to leave England. But surely it was more than that because, compared to many, these families lived well.

The Clipson family story and later newspaper accounts mention that the Clipsons possessed watches, rings, and other jewelry they had brought from England and passed

Gathering for the Grand Prairie

down to their children. A late nineteenth-century account from a Catlin newspaper column describes furniture and the oil portraits of Will Clipson and his second wife, Matilda. They brought these with them to America. A Clipson descendant in Indiana displays those portraits in her dining room today.

A diary covering the late 1840s kept by Eliza Jones describes a family of women with time on their hands to go visiting and seeing the sights. When they finally set out for America, many of the Grand Prairie families traveled first class. Their possessions, habits, and choices suggest an amount of discretionary income that would have made life in England relatively easy.

In their working-class neighborhoods, though, diseases and people living on the edge surrounded them. Crowded conditions made it hard even for the relatively prosperous to find housing. After all, the future Grand Prairie immigrants, like most Londoners, spent much of the 1840s moving around, reflecting either problems or aspirations.

An attraction to America's Midwest for the Grand Prairie English might have been a romantic notion of an agricultural life brought on by watching such English landowners as Sir Joseph Banks in Lincolnshire. John Kenneth Galbraith described another attitude toward rural life as it applied to immigrant farmers in Ontario in his book *The Scotch*. "The Scotch believed, I have always thought rightly, that they were [superior]. They considered agriculture an inherently superior vocation. It placed a man in his fit relation to nature; it abjured the artificialities of urban existence. It gave him peace and independence. It was morally superior for it required manual labor ... none of them believed that clerking in a store or weighing in at an elevator was work."

It's not clear how much manual labor the Grand Prairie English anticipated, but, given their haste to acquire land, the moral superiority of working it might have been a factor in their decisions. Practicing agriculture like Banks and other English landowners did would confer the kind of status and wealth that were hard to achieve when working in town for a living. This would have been true even for such successful entrepreneurs as Henry Jones.

There was, and still is, a mindset that also deeply influenced any English person's decision to move. It is different from the mindset held by this nation of immigrants called the United States. England has been for centuries a nation of emigrants. That history, accompanied by another typical English behavior—that of facing adventure on the high seas—made it easier to leave and take up a new life elsewhere. The Church family, whose members emigrated to Australia, North America, Brazil, and the East Indies, was the best example among the Grand Prairie English of this attitude toward leaving and adventure.

Dickens addressed this facet of English behavior in *David Copperfield*. Mr. Peggoty and Emily plan to immigrate to Australia. "There's mighty countries fur from heer. Our future life lays over the sea," Peggoty said in the story. Mr. Micawber also immigrated to Australia, saying, "It was the dream of my youth."

"No better opening anywhere," said David's aunt in support of the decision, "for a man who conducts himself well and is industrious."

England remains a nation of emigrants. The *Daily Mail* reported on September 16, 2008, that "record numbers are leaving the UK to settle abroad. In the 12 months to [this past] June, 406,000 left to live overseas—the highest level since records began. Australia, Spain, America, New

Gathering for the Grand Prairie

Zealand, and France are among the most popular destinations. And the most recent study by development company Property International found that 50 percent of us would consider moving overseas."

No records hint as to when the Grand Prairie group began thinking of emigrating. But those that left in 1849 and 1850—the Joneses, Churches, Puzeys, and their friends and family members—would have spent the 1840s saving their money and solidifying their plans.

—⁊⁊⁊—

In June 1841, England's census found that London's population had doubled since 1801, the date of the first rudimentary civil census, with more than 2.2 million inhabitants. The 1841 census reveals that several of the Grand Prairie immigrants were living within shouting distance of one another.

Henry and Sarah Jones had moved south of the Thames to Lambeth and were living with their six children in the Henry Cottages along the Grand Surrey Canal. Henry's occupation was recorded as a gas fitter. For what apparently would be a growing relationship between Henry and his wife's niece, Eliza Hough, she and her new husband, Thomas Bentley, conveniently lived next door. Thomas's three children by his first wife, the late Sally Holland, were living with them. So were his younger brother James and their father, Thomas. Their father, Thomas, died in 1843, but James would accompany his brother's family to the Grand Prairie.

The Clipsons were still moving around from house to house but were also south of the Thames. Small row houses near the more elaborate Brunswick House (still standing) characterized their 1841 address, Belmont Place.

Sophia Puzey Church and her family were living north of the Thames at their longtime address of Number 10 Mile End Road in Stepney. About the same time, the younger Henry Puzey, the nephew of Sophia Church and Richard Puzey, who had remained in Berkshire, moved to London and became apprenticed to his uncles, Henry and Thomas, at their dry goods stores. The family story says he was there for five years with his father, Joseph, paying forty pounds sterling as "tuition." Henry stayed on for four more years as a paid employee at a rate of twenty to forty pounds, depending on the year, according to the family history.

—᙮᙮᙮—

In 1843 London was full of news that the future immigrants to the Grand Prairie would have talked about. In January Sir Robert Peel's private secretary, Edmund Drummond, was murdered on his way to Downing Street. The culprit was a delusional man, Daniel McNaughton, who thought Peel was out to get him. The subsequent trial established that McNaughton was not guilty by reason of insanity and created the well-known legal insanity defense known as the McNaughton rule. This generally states that a person must be able to understand right and wrong to be held responsible for a crime.

Soon after the murder, the Grand Prairie Londoners were treated to a new way to cross the river. In March the Thames tunnel opened. In the works for eighteen years, it was the first tunnel to be built under a navigable river and was hailed as the eighth wonder of the world. Marc Brunel designed it. Shops lined its 1,506 feet. But the owners had neglected to build ramps for horses and wagons, so only

Gathering for the Grand Prairie

pedestrians could use it. It cost one penny to cross. In time it became a popular site for robbery and prostitution. Now it is part of the London Underground.

—∞—

At some point in the 1840s, one future immigrant attended a lecture in London extolling the advantages of making a life in east-central Illinois. It was all about land.

The young Henry Puzey would have heard that settlers had been filling southern Illinois since 1800 in the "American Bottom," which lay amid the Wabash, Ohio, and Mississippi Rivers. Other parts of the state, however, especially the east-central portion called Vermilion County, were beckoning now that Illinois was no longer a wild frontier.

The Indian tribes had been moved out, so they were no longer threats. While the buffalo were also gone, the fields and forests yielded plenty of game and edible wild plants such as sassafras root, mushrooms, berries, and greens.

The state was growing fast. Illinois' 1840 census showed a population of almost 480,000, fourteenth among the twenty-six states. It contained 35,235,299 acres, more than Ohio, Indiana, or Iowa, which, unlike Illinois, was only partially surveyed. Land was still available in the central part of the state, but it would soon be taken up. The lecture's message: act fast or miss out.

The prairie land was fertile and drawn into sections that made for reliable buying and selling. It was also cheap—$1.25 an acre if one avoided the speculators and bought it from the federal government. Taxes were low. The growing season was long. The extensive river system that wove through the middle of the continent made it easy

87

to ship grain and cattle to market. A young English person would have to work hard, but by owning land, he or she could break free of the constraints the homeland imposed.

Hawthorne, influenced by his patriotism, compared an American's control over personal fate with an English person's. "The public life of America is lived through the mind and heart of every man in it. [In England] the people feel they have nothing to do with what is going forward, and I suspect care little or nothing about it. Such things they permit to be the exclusive affair of the higher classes." Hawthorne's judgment about the English, however, neglects to account for the popularity of the dozens of newspapers that provided detailed information about public affairs as well as private sensations.

In any case, one gets the idea that such a well-educated, confident, and energetic young man as Henry Puzey would fester under a rigid hierarchical system if he could imagine a better kind. In America he could rely on his own hard work and not on someone else for his success. He would become admired and emulated. He would be free of a landlord and an employer, even one such as his uncle. Religious institutions, not linked with government, couldn't impose taxes on those living within their spheres of influence. A young man could vote. He could help shape a community still unformed, but bursting with excitement over its possibilities.

It was true the land was a bit soggy. The English, though, as everyone knew, were experts at drainage and were already accomplishing that successfully on the prairie. Where the sod was not broken, a farmer could turn out his cattle, sheep, and hogs to graze on the nutritious prairie grass, and the animals would almost take care of themselves.

The federal land office for east-central Illinois was conveniently located in Danville, the largest town in Vermilion

Gathering for the Grand Prairie

County. So no travel time need be wasted in buying land. The American settlers in Illinois spoke the same language as Henry, and a good number of their ancestors had come from England too, albeit a century or two earlier. Moreover, these farmers, no matter where they came from, regarded themselves as landowners and gentry—not peasants as European farmers viewed themselves. Henry would be surrounded with neighbors with fine aspirations and model standards of behavior.

One didn't even have to farm to succeed. Since Henry had apprenticed in the dry goods industry, that would have been important to him. The rapidly growing number of farmers coming from Kentucky, Indiana, New York, and states in New England needed mills, shops, goods, and services that such well-trained men as Henry Puzey could provide.

Furthermore, Illinois was relatively free of the squabbles that affected other states where land titles were unclear and attitudes toward slavery bitterly divided neighbors. English men and women by this time strongly disapproved of slavery, so Illinois' slave-free status meant they didn't have to confront that disgrace.

Steamboats had recently made shipping stock and grain quicker and cheaper, and with an extensive network of canals, traveling between Illinois and East Coast ports had become easier. Railroads were planned that would make getting goods to market even more efficient.

—ɯ—

Henry Puzey's 1922 obituary and an earlier biographical sketch claimed it was through this lecture, sponsored by Isaac Sandusky, that he heard about the marvels of

89

east-central Illinois. Isaac's family, known by both the Sandusky and Sodowsky surnames, had settled in Vermilion County late in the 1820s and had become some of its largest landowners. Isaac, a land speculator, was clever in sponsoring the lecture. The more people he persuaded to immigrate to his section of the Grand Prairie, the more potential buyers he would have for his extensive land holdings when he decided to sell.

Henry was probably not the only future Illinois immigrant at the lecture, but one wouldn't have had to go to that lecture to learn about the promise of Illinois. Both Americans and English visitors wrote numerous books in circulation in England that detailed conditions in the state and gave advice. Throughout the 1840s and beyond London newspapers carried articles about emigration. They recommended how emigrants should proceed and where they should go. Illinois was at the top of everyone's list. In 1849 a writer with the pen name "Backwoods" regularly gave advice in the newspapers. He wrote: "To the man who has a large family … and who wishes to see his children placed in a position which they can never attain in this country, that being 'lords of the soil,'… if he can command a small capital, his position will, with ordinary prudence, be greatly bettered."

Wealth was almost guaranteed. Life was described in idyllic terms, much exaggerated. "Almost every man takes a newspaper," extolled one writer, and "in western cities all the necessaries and luxuries can be obtained."

—∞—

Isaac Sandusky's lecture might have pointed out another attraction of Vermilion County. English-born men and women were already living there.

Gathering for the Grand Prairie

In the 1830s and early 1840s, Abraham Mann, the Reverend William Cork, and their friends and relatives had immigrated to an area that became Ross Township in the northern part of the county. There were about seventeen households in all. The wealthy Mann was from Bedfordshire. The Corks were from Devonshire.

Abraham Mann's father was one of the first Englishmen to explore central Illinois with an eye toward settlement. John Mann was a salesman for a paint and oil maker in London before the American Revolution. His territory began in New Orleans, and his task was to extend it up the Mississippi as settlement increased.

John Mann used his time as a salesman to scout out places suitable for farming. He apparently suffered a financial loss from the American Revolution because the newly formed U.S. government gave him thirty thousand acres in New York and Louisiana as compensation, according to a history of Mann's Chapel, the small church building Abraham had built with his own bricks. While those acres weren't worth much at the time, their value grew, adding to John Mann's and later his son's wealth.

John Mann's journeys had taken him to Illinois, the beauty and promise of which he conveyed to his son Abraham. In the early 1830s, Abraham led his family to Vermilion County, bought a section of land, and proceeded to raise cattle. At the end of his life, his farm comprised about five thousand acres.

Abraham Mann was a benefactor in his adopted community and an enthusiastic employer of newly arrived immigrants. In 1850, according to the census, five of his eight farmworkers were English-born.

One gets the idea that Mann had many connections both in England and Vermilion County. Mann returned

Getting to Grand Prairie

to his birthplace in Leighton Buzzard, Bedfordshire, to live at least once before returning to settle permanently in Ross Township.

Richard Puzey, the first person to emigrate from the families whose stories this work follows, was quick to make a connection with the Ross Township English after he arrived in the Grand Prairie. Within two years he had married Amelia Cork, the sister of Mann's friend, the Reverend William Cork.

11

A Girl's Life in London

In 1846 Henry and Sarah Jones's daughter Emily, also called Emma, gave her seventeen-year-old sister Eliza a diary. Someone saved it. Someone else transcribed it and gave the transcription to the historical society in the Illinois village where Eliza spent most of her life.

Eliza is an economical writer and, frustratingly, doesn't address certain important conversations her family must have been having given that, by the end of the decade, they had moved to another continent. She jotted down a sentence or two every so often and just hit the highlights. Eliza was formal. In the diary she called her mother's niece, her own cousin, "Mrs. Bentley."

She must have endured boredom after her arrival in the Grand Prairie, and she later faced tragedy. In her diary, however, the appealing Eliza comes to life in a glow of activity and happiness. Why shouldn't she be happy? She had close sisters, good friends, and indulgent parents. One of her first entries tells us the family has moved again, this time to Oxford Terrace on October 16, 1846. She also found love. In one of her early entries, on April 1, 1847, she wrote, "Introduced to Mr. Browne."

William Browne was an ideal suitor—attentive, kind, and thoughtful, coming up with a brooch, a box of "plumes," a collar, a "patchwork," and presumably other gifts now and then. He took Eliza for tea at the Dartmouth Arms at Sydenham, a historic pub in a romantic canal-side setting. He escorted her to the Coliseum and the Cyclorama. He accompanied her and her sister Emily to a thrilling melodrama at the Victoria Theatre, located near the south bank of the Thames where the London Eye now rotates. Coincidentally it was the girls' father's company that had lit the theater and the Coliseum with gas lights.

On October 17, 1848, Eliza noted that she and her cousin Mary Ann Hough had tea at the Clipsons', which proves the London connection between these Grand Prairie families. The ever-reliable William "fetched us home."

Another entry suggests an additional connection. In October 1848, Eliza "went to tea with Emma Horniblow," and William accompanied her home once again. In the 1850 census of Illinois, English-born Edwin Horniblow was recorded as living with another London couple who had accompanied the Joneses on their migration. The Horniblows don't figure greatly in the Grand Prairie settlement, but this is further evidence of the many London connections the immigrants enjoyed with one another before arriving in Illinois.

Eliza's diary shows she was constantly on the move. She doesn't always say with whom she went on her excursions, but it is apparent she explored London with determination and delight. She went to Anerley in southeast London, most likely for a visit to the woodlands around Penge Place. She enjoyed the St. Helena Tea Gardens at Rotherhithe to the east and to the west, Hampton Court, the palace long associated with the British monarchs. She "had tea

A Girl's Life in London

at Greenwich," the enthrallingly scientific city where the world's longitude begins. She traveled frequently by river steamer, since most of the places she mentions are located along the Thames.

In July 1848, she visited the Surrey Zoological Gardens where spectacles such as the "eruption of Vesuvius" could be seen. Her father had created the illuminated Gothic Bridge, "its contours lined with hundreds of gas light buckets," according to the *London Illustrated News*.

Eliza's diary detailed her love of the theater. She patronized the Surrey, where she was "highly delighted with Phelps," a noted tragedian. In December 1848, she went to the Sadler's Wells Theatre, which specialized in Shakespeare. William Browne took her to the Adelphi Theatre on the Strand, which featured comedy, musical theater, and sometimes Charles Dickens's stories adapted into plays.

Eliza also described her handiwork—her father's pencil case, and for her cousin Eliza's husband, Thomas Bentley, a purse she began on December 30, 1847. She completed it six days later. She did not forget her young man, William, whose purse she completed on October 31, 1848, having started it on October 24.

Tantalizing phrases make a reader want to know more. "Received a paper from America," was written on November 11, 1847. What was in it, and who was it from? A month later she wrote, "Received a letter from Uncle." Which uncle, where was he, and what did the letter say? "William Allison left London on 19 February 1848." Who was he? Was he also going to America?

Eliza noted life's passages. Her cousin Samuel Hough died at age nineteen on December 4, 1847. The Clipsons' daughter Matilda, who died three years later, was christened on December 26, 1847. Arthur, Eliza's youngest brother and

95

the last child born to Henry and Sarah Jones, was born at one o'clock on July 14, 1848. About a month later, Eliza wrote of the christening of Henry and Emily Bentley. They were born a couple years apart but christened together. When the Jones and Bentley families arrived in New York about a year later, however, only Henry was with them.

In January 1849 she noted that Frederick Clipson was christened at St. Mary's Newington. He would die the next September after Eliza had emigrated.

She mentioned her mother only twice in her diary—when she and Mrs. Bentley went visiting, and when her "mother and father dined at Mr. Bentley's." That was only four months before her father left for America. But her father and some of his business activities were chronicled. Her father's trip on September 3, 1848, to Devonshire with Mr. Lee, whom Henry mentioned later in a letter written aboard his immigrant ship, got her attention. Less than two weeks later, she recorded that her father "opened the shop in 6 Rose King Street, [Covent] Garden." This is a puzzling notation since he had earlier been at the same address. Henry gave Eliza a book on November 29, 1848, but she did not feel obliged to note the title. A friend or business associate whom Henry mentioned in his shipboard letter appeared at their London house in December 1848. "The three Mr. Lockes and Mrs. Locke came to tea with us," wrote Eliza. Eliza employed both the name King Street and Rose Street as the address of her father's business. When she visited King Street, William escorted her home. At the end of January 1849, her brother Richard, who by then was working for their father, went to Cambridge and returned the next day. With that speed he must have traveled by train. One of her final entries, also at the end of January 1849, is

A Girl's Life in London

the only clue to their future plans. "Father sold the business in King Street."

The diary confirms that the Bentley and Jones families spent a lot of time together on excursions. They visited one another's houses, celebrated with parties, and enjoyed a twelfth-cake together a few days after the Twelfth Day of Christmas.

Eliza didn't mention events outside her family that were unsettling London—another revolution in France in February 1848, working-class riots in the Kennington Common, or another cholera epidemic that killed fifteen thousand Londoners in 1848.

Except for the sale of H. Jones & Co. Gas Fitters early in 1849, Eliza never mentioned any plans for leaving London, her feelings about the upcoming adventure, steps other families were taking in preparation for their journeys, or her hopes for their new lives in America.

One of the last entries before she left London occurred in February 1849: "I had a new pair of boots." With all her excursions, she must have needed them badly.

97

12

Tossed at Sea

In 1847 James Knox Polk was the president of the United States, Augustus C. French was the governor of Illinois, and Richard Puzey became the first of his group to leave England for the Grand Prairie. He is said to have taken with him at least two books. One was his wife's Bible. The other was the ever-popular *The Saints Everlasting Rest* by Richard Baxter, published in 1822. According to his granddaughter Matilda, Richard's version was complete with 616 pages and beautifully etched illustrations. He might have taken plants, especially gooseberries. The American varieties were believed to be less tasty than the English ones, and descriptions of his farm confirm he planted them on his property.

Shops kept lists of articles emigrants would need either on the ship or when they arrived. English farm implements were not appropriate for breaking tough prairie grass roots. One ironmonger on Tower Hill, however, recommended that, in addition to firearms, emigrants might want to take cornmills, ploughs, wheelbarrows, and other tools. Apparently some of the English believed these items were unavailable in America.

99

Getting to Grand Prairie

Richard had no family member with him when he emigrated, but he wasn't alone. While the number of British immigrants to America did not equal the number of Irish or Germans, it was nonetheless substantial. Four and a half million English, Scottish, and Welsh immigrants arrived in America between 1820 and 1930. They arrived in three waves with the largest, about a half a million, coming in the mid-1840s to the mid-1850s, the decade in which the Puzeys, Joneses, and their friends came. Historian Charlotte Erickson claimed the time period from 1850 to 1880 contained "the greatest movement of people the world has ever known."

Nonconformists were more likely to immigrate to America than Church of England members. Although the Grand Prairie English appear to have been motivated by reasons other than religion, Richard Puzey might have had Methodist leanings that would have supported the notion to leave. He did later found a Methodist Church. He is not, however, listed in Berkshire County's list of nonconformists at the time. The story of his devotion to Methodism prior to his move to America is written by descendants who were devout Methodists. One should, therefore, take such assertions with a grain of salt.

Another possible reason for Richard Puzey's early immigration was the loss of his wife and perhaps children. He simply had little reason to remain in England. Furthermore two of his older brothers were already in America, having left in the previous decade. Their new homes in Ontario and New York State did not appeal to him, since land was already taken up in those locations. Besides, Richard had a place to stay with Isaac Sandusky once he got to Illinois. As a single man, he didn't have to save as many pounds to pay for passage as did men with large families.

100

The Vandersteens, two Swannell brothers, the Kays, and the Onleys (also called the Olneys) were the next to arrive. George and Jane Vandersteen disembarked from the *American Eagle* with their infant son, George, on November 3, 1848. George reported his arrival year on the 1900 census when he was living as a widower with his daughter, Ellen Phettiplace, in Mississippi. They came with Jane's mother, Mary Ann Burney, and her three sisters, Mary Ann, Ellen, and Sarah Burney, but the women did not stay long. They had returned to England by the mid-1850s.

George Vandersteen was a musician and a cabinetmaker, but in the 1850 Vermilion County census, he listed his occupation as a miller. Like many men of that time, he probably worked at whatever job he could find. Frederick and John Swannell also arrived the same year on the *Richard Cobden.*

It is hard to know what to do with the Vandersteens, Swannells, Kays, and Onleys in this story. They were not mentioned in Eliza's diary, Henry's letter, or the Church family's later letters. They can't, therefore, be connected in London with any certainty to the Joneses, Puzeys, Bentleys, Churches, Clipsons, or Dickinsons.

Their arrival, however, is hard to attribute only to coincidence. These families had lived in London in the same neighborhoods as the connected ones did. The Onleys purchased land next to Richard Puzey shortly after their arrival. Some Puzey and Church descendants were left with the impression these families were part of their immigrant group. There seemed to be unusual trust and affection among the families. George Vandersteen witnessed Henry Jones's 1857 will, and the Vandersteens and Kays were warmly welcomed at "English reunions" later in the century. For a Londoner, Vermilion County, Illinois, is a hard-to-find place across a treacherous sea and a third of the way across an unfamiliar continent. Perhaps members of these families heard Isaac Sandusky's lecture at the same time Henry Puzey did.

Family memories, however, are not always accurate. The feelings of affection could have developed in America and not London. With incomplete information, all that is known is that these London families arrived in the same years as the families that are proven to be connected in London.

Another difficulty of confidently fitting them in this story is that the Swannells, Onleys, and Kays left fewer records in Vermilion County because some descendants moved away and lost touch with those who stayed.

The Craddocks and the Milemores (or Millimores) are also unaccounted for. The Craddocks settled in southern Vermilion County about the same time as the interconnected families. While some present-day descendants of the Grand Prairie English believe they were part of the group, there is so far no evidence they were acquainted in England with the other immigrants. They can't be located in London, Berkshire, or Lincolnshire, and there is no mention of them in any of the primary families' papers. English immigrant William Millimore was at Henry Jones's bedside when he died. James Milemore arrived in 1852 from Bedfordshire, the home of the early immigrant Abraham Mann. Several families with the last name of Swannell lived in the same village as the Milemores in Bedfordshire. Without further study, however, it is unclear how closely related, if at all, those Swannells were to the Swannell family who immigrated to the Grand Prairie. No evidence so far exists of a friendship the Joneses, Puzeys, or the other families with proven connections had with the Craddocks or Milemores in England.

Tossed at Sea

—ᴍ—

The next Englishman to arrive was Henry Church's eighteen-year-old nephew, George William Frederick Church, Gentleman, stepping off the *Devonshire* in New York on July 15, 1848, with no one that can be identified. He did not, however, go immediately to the Grand Prairie. First he went to upstate New York, where he lived with his Uncle William for a year and four months, and then left for Illinois. His father is said to have paid William twenty dollars a month for George's keep and instruction in farming.

George W. F. Church was well-educated, so it is interesting that George and eventually his brothers, Albert and Adolphus, emigrated. Their father, George Zephaniah Church, the prosperous clerk at the Bank of England, might have wanted his sons to follow in his footsteps. But that wasn't the case. The family's history showed that emigration could also bring great wealth. As part of the East India Company, Thomas, George Zephaniah's brother, illustrated this well.

The last English immigrants to arrive in Vermilion County in 1848 were the rest of the Swannells. They disembarked in New York from the *Mediator* on September 1. Sarah Swannell, a forty-four-year-old widow and the stepmother of John and Frederick Swannell, led a party of six: her children, Alfred, Henry, Eliza, and Maria, her stepson William, and her older sister Mary Lound. Sarah had become a member of a Scotch Presbyterian Church in London. Her family would have an easier time as dissenters in America.

In 1849 the population of this bit of the Grand Prairie increased by at least forty-one English men, women, and children. Total immigration from the United Kingdom to the United States in that year was 219,450. The destination with the next highest numbers of British immigrants was Canada with 41,367 people entering.

Some immigrants were headed for California, where gold nuggets had been found at Sutter's Mill. But the Grand Prairie immigrants were drawn by the tangible asset of land, not the fantasy of finding gold. The men of these families had no get-rich-quick dreams. Their goal was to establish prosperous domestic lives, and they were willing to work hard for it. Their dream was to build a future with status, independence, and wealth in a community that would nourish them. Immigrating together to the same place was a way to continue their familiar English ways while enjoying the benefits their new community was bound to offer.

The next group, thirteen members of the Onley family and their relatives, the Ansteads or Ansteads, arrived in New York on the *Westminster* on April 6, 1849. Henry Church came without his family in that spring also we presume. Philip Pusey, Richard and Sophia's half brother, arrived at some point in 1849, according to his account at the Old Settlers' Meeting in the late nineteenth century and also the information he supplied when applying for citizenship. He cannot be found in the 1850 in the U.S. census or in England's 1851 census, though, in any location.

John and Sarah Taylor arrived from England next. They knew the Jones family through John's sister Caroline, who had married Henry Jones's friend and immigration companion Henry Shale in 1833. The Taylors arrived in New York on April 30, 1849, on the *Independence*. John was a blacksmith, and in 1841 he, Sarah, and their five children had lived in the parish of St. George the Martyr in Southwark near other Grand Prairie families. When John and Sarah got off the ship in New York, though, no children were with them. The most likely explanation is that the children died in the epidemics that hit London throughout the 1840s. As two more people with nothing left to lose, the idea of emigrating with friends to a new land must have been appealing. With no descendants to preserve their story, though, little is known about them.

The irrepressible forty-four-year-old Henry Jones boarded the American-built *Hendrik Hudson* on a first-class ticket on May 3, 1849, when it stopped in Portsmouth to pick up more passengers. His friends Henry Shale, forty-five, and Thomas Hind, thirty-four, accompanied him. It took the Pilgrims 101 days to sail from Southampton, England, to the Massachusetts coast in 1620. Henry Jones made the crossing to America in thirty-five days.

Henry's son Richard and Mr. Locke, whom Eliza had mentioned in her diary, left Henry at Portsmouth at eleven o'clock in the morning and returned to London. They had stayed overnight at the Quebec Hotel, and then they said their good-byes. Henry left for the ship with Hind and the ship's American captain, Isaiah Pratt. By one o'clock, the ship was underway.

Henry kept a diary of his voyage that became a letter he sent to his wife and children from New York after he arrived. The family kept the letter and included a transcription in the material they gave to the Catlin Historical Society. The feeling that comes through most clearly in his diary is Henry's enjoyment of the trip, despite bad weather.

He noted the beauty of the sea and the sunsets. He enjoyed stout, gin, and brandy throughout the day. He played checkers with Hind and Shale and had meals and good conversations with the captain. He usually went to bed in his cabin around ten or eleven o'clock and woke around

Getting to Grand Prairie

six or seven in the morning. The other passengers fascinated him, and he spent time with many of them. On the first day out, he reported: "During the afternoon I had conversation with a Boston gentleman who has been in China five years and a half. He told me if I wished to speculate in the leather trade that I should find Lynn near Boston a very good place."

This recommendation was not enough to persuade Henry to alter his plans for the Grand Prairie. If an immigrant had wanted to buy land in the mid-nineteenth century, he or she would not have chosen New England. New Englanders themselves were abandoning that region's rocky fields and short growing season for the deep soil and long summers in what they called the Northwest.

On May 5, Henry saw Land's End on the coast of Cornwall. He gave no indication he felt any nostalgia at his last glimpse of England. Instead he was delighted when the ship began to make good time with a strong wind. While the crew coped with rough weather and other passengers succumbed to seasickness, Henry, Henry Shale, and Thomas Hind toasted their friends and family members back in England.

Other passengers were exotic world travelers, and they passed on information about parts of the world the Grand Prairie immigrants would never see. "Mr. Hind's chum is a West Indian Jew, a lad about fourteen years old...[he] told us that in Jamaica the sun sometimes is so hot that people are sun-stroke to death the same way you would be by lightning."

A four-year-old child died on May 11 and was "consigned to a watery grave."

Bad weather afflicted the ship, and Henry reported that "the second-class and steerage passengers were kept under hatches, and at two o'clock the alarm was great. In the saloons

106

the utmost confusion prevailed. One lady was thrown from her chair, hurt her side, and fainted. On bringing her to she made anxious inquiries after her comb, which, by the bye, cost eight pounds, thinking more of that than the danger she was in. At ten o'clock in the evening the storm was still raging. The passengers laying about the saloons, a few in their berths with their clothes on, and by twelve the utmost alarm prevailing."

By noon on May 13, they had passed through that storm, and Henry was able to enjoy an excellent meal and good conversation with his friends and the captain. He could endure the tossing of the waves, but he recommended his family bring medicine for constipation, which he, Hind, and Shale wished they had done.

On May 15, another storm tore off the foremast's topsail, which blew into the water and was lost. Henry reported the incident but claimed that the summer crossing his family intended to undertake would enjoy better weather. He seemed determined not to frighten Sarah and the others who would come after him, even while he noted the tossing sea.

One of the first-class passengers, sixty-four-year-old Mary Hone, was reassuring about Henry's family's crossing. "We had an old lady," he wrote, "who had made three voyages across the Atlantic within the last four years, and she told me that this was the sorest passage she had ever made."

On May 16, Henry played "shuffle" on the deck and talked with a woman in second class. The storm had damaged her clothes and the food she had brought to eat on the voyage. The next night Thomas Hind's berth broke, and his cabin flooded. Henry's cabin was spared.

Storms continued to wrack the *Hendrik Hudson*, but Henry was usually able to get a good night's rest. On May 21, he reported that, for the first time since boarding the

Getting to Grand Prairie

ship, he had dinner "without any of the things on the table being upset." The passengers had milk from a cow kept on board. Henry reported that the few English who traveled as second-class and steerage passengers complained that the Germans monopolized the cooking fire and utensils.

The weather turned ugly by nighttime, and Henry was kept awake. "One of the yards broke and the other damaged...the Captain told me that he never knew so rough a May. At seven o'clock the ship received such a shock from the waves that the man at the wheel was completely thrown over it. The Captain, hearing the noise, rushed on deck, caught the helm, and guided the vessel; the poor fellow was very much shook. The evening turned out fine, and we had been going four and a half knots all day. I being much fatigued with the late tempest turned in by nine o'clock."

Henry was encouraged by one event. "Having recorded the death of the child on the ship, I have now to do the same to the birth of one who was born in steerage last night. It appears that young newcomer is a fine boy, and by the accounts that I have gathered afterwards is the pride of his mother's heart."

When Henry turned forty-five on May 25, the ship was off Newfoundland in dense fog. Hind and Shale made sure their friend had his favorite gooseberry pie as a treat at dinner, and Henry wrote that he hoped his family was enjoying one too in honor of the occasion.

Henry justified his first-class cabin to his family. "In conversation with the Captain, he said if persons wished to be thought anything of in the new world, they would not only travel first class but make a good appearance, for the Americans think much of these things."

The ship carried two hundred fifty second-class and steerage passengers, according to Henry, and the first-class

108

passengers and crew together numbered about fifty. His numbers, however, differed from the captain's figures. Captain Pratt counted three hundred fifty total passengers—forty-four first-class passengers with two infants, twenty-two second-class passengers and 267 passengers in steerage with fifteen infants. Henry described the steerage compartment as hot, crowded, and afflicted by fever. Although most passengers had to carry their own food and cook it, the first-class passengers enjoyed roast beef and plum pudding at the captain's table.

Perhaps Eliza Bentley was worried about conditions on the ship because Henry instructed his family to tell her how well the first-class passengers lived. He was entertained by sighting several whales and pondering the four-and-a-half-hour time difference between his location and London.

The only regret Henry Jones expressed about leaving London was near the end of the voyage when he, Shale, and Hind consoled one another about not being able to meet Will Clipson at Peckham. It is interesting to speculate what they might have done in that growing London neighborhood south of the Thames.

The weather was good as the ship approached New York, and Henry reported seeing a school of mackerel "by the thousands." He was up early on June 7 to watch a steamer tow them into New York Harbor. Doctors came on board to examine the second-class and steerage passengers, who were all declared healthy. The first-class passengers were not examined but left to their breakfast as the doctors peered into the dining room. The passengers weren't finished with examinations, though. Because of that Henry had instructions for his wife about their shipping boxes.

"By eight o'clock the custom house officers came on board and examined our luggage. If I may so term it, they

Getting to Grand Prairie

[were the most] gentlemanly set of men I ever met with. They merely caused your boxes to be opened, and without disturbing anything we closed them again. I could not get my portmanteau open, and they even passed that. Their courtesy induced me to ask them respecting furniture and they told me, on your own, you would not have to pay duty if you brought it with you. Therefore if you have not parted with the looking glasses, bedsteads or any other articles that you would value you may bring them with you, and anything that you cannot part with before you leave you had better get my trustees to do it for you after you have left."

By noon the passengers could leave the ship. Thomas Hind eventually returned to England. Henry Shale can't be found after this voyage. Henry Jones doesn't mention them after he landed. His first duty was to go to Jersey City to track down a Mr. Bridgett, who helped immigrants with arrangements. Shortly thereafter he retrieved mail that must have come on the faster mail packets that were operating in the Atlantic. He had letters from his son Richard, his wife's relative Mr. Hough, and his friends Locke, Lea, and Will Clipson.

His comments about Richard's letter lead one to believe there was talk in London about his departure. "The false assertions made against me by Richard's statement in his letter, respecting my leaving England because I could not pay my way, I scarcely need say to you it is entirely false, for I believe all who know me [know] that I am incapable of being guilty of such a dishonorable act, and the only way to treat such scoundrels is with silent contempt." Richard's letter did not survive, so it isn't clear who made false statements or why.

Henry praised New York's prosperity and declared that Brooklyn and Jersey City would be beautiful places in which to live. Then he told his family to address a letter to him in care of Mr. Bridgett, telling him when they would arrive and the name of their ship. Henry was to leave for St. Louis the next day, perhaps to visit his brother-in-law, Samuel Hough, and maybe even to nip over to Vermilion County. He intended to return to New York in time to meet their ship.

His advice was authoritative for a man who had just landed for the first time in a foreign country. "You need not land your goods till you get a place to settle in so don't encumber yourself in first starting from the ship. Be sure you have your names put clear on each of your boxes and also firm and bring an inventory of the contents. Tell Richard I know of nothing that he can bring with him more than I have named, but I should like to have a statement of my financial affairs from my trustees. Let him not forget it."

In case something happened, and he wasn't there to meet them, he told them to contact Mr. Bridgett. Then he detailed the route they were to take. "The Railway Second-Class and Steamboat expenses from here to St. Louis is eighteen dollars, and that is the quick way going, and it takes six days without stoppage, but you can go another route for nine dollars, but that takes 16 days. In both cases in addition you have to find your living, etc."

Henry's letter confirms that the immigrants preferred rail and river travel to the other means of getting to the Grand Prairie. The best route would include trains from New York to meet the Ohio River steamers, which were faster than canal boats and more comfortable than bouncy stagecoaches. But the Jones family would have had to rely on stagecoach lines or wagons in getting to Vermilion County from St. Louis or from other boat landings, since rail lines had yet to be built in central Illinois. By 1836 a rudimentary post road crossed the state between the Mississippi

111

Getting to Grand Prairie

River on the west and Danville on the east, going through Springfield, Decatur, and Urbana. If any immigrant came from New York via the Erie Canal and the Great Lakes, he or she would have taken a road improved in 1849. It generally followed the 1820s trading post route, Hubbard's Trace, going south from Chicago to Danville. Illinois Route 1, also known as the Dixie Highway, now roughly follows that historic path. Immigrants could also get to central Illinois on the Erie Canal and then to other canals that crossed Ohio and Indiana. If they got by rail to a packet boat on the Ohio River, they could have poled up the Wabash to Vincennes or even Covington, Indiana. From there they could have taken wagons or stagecoaches to Vermilion County along roads that were increasingly being built or improved. Except for Henry Jones, none of the Grand Prairie English saw fit to pass on to the next generations how they actually got from New York Harbor to the Grand Prairie, but it must have been a laborious, sweaty, exciting journey.

In his letter home, Henry did not forget the rest of his family and friends. "Remember me kindly to Mr. Bentley not forgetting Mrs. B....Also don't forget me to Mr. Hind and his good lady and then one and all of you have nothing to fear but a great deal to hope for. Still I have a good opinion of everything and something can be done and a few months will see us I hope at work in right earnest and to our material benefit."

Henry ends his letter by telling his family how much he misses them and why they are emigrating. "Believe me," he wrote, "it has been a hard trial to be parted from you all so long, and no doubt you have felt the same. I hope it will be for a short duration and, my children, for your sake alone have I crossed the Atlantic to place you all if possible hereafter in a comfortable position where you may be able to settle down in peace and happiness and beyond the reach of want."

Henry was to achieve his goal. He chose land acquisition in a farming community rather than plying his gas fitting trade in a city as the means of realizing his ambitions. His ending statement, though, shows that Henry's dreams were essentially the same as all immigrants who have come to America.

—ᴀᴧ—

The Jones family story says that Sarah Jones and her entourage left England on July 14, 1849. On August 18, 1849, they were among the first passengers recorded on the *Northumberland* upon landing in New York. What an entourage it was—twenty-five people in all. Not only had Sarah brought her seven children—Richard, Sarah, Eliza, Emily, Louisa, Frederick, and Arthur—but also her seventy-five-year-old mother, Ann Hough, and her nieces, Mary Ann, aged twenty-one, and Elizabeth Hough, seventeen. Those girls were the daughters of the Samuel Hough in St. Louis and the sisters of the Samuel Hough whose death at age nineteen in 1847 had been recorded in Eliza Jones's diary. Mary Ann and Elizabeth eventually joined their father in St. Louis.

Also on the ship were Sarah Jones's niece, Eliza Hough Bentley, and her husband, Thomas, their two surviving children, Alice, aged five, and Henry, an infant, and Thomas Bentley's three children by his first wife—Thomas, Sarah, and Fanny. Finally, there was the elder Thomas's brother James.

Sarah Jones had friends with her too. William and Elizabeth Hind were an older childless couple who had been neighbors of the Joneses and Bentleys. William, a printer and compositor, was the older brother of Henry's shipmate,

113

Thomas Hind. James Thompson, a Scotsman, his English wife, Ann, and their infant daughter, Ann, were listed on the ship near the Jones family. They turned up later living two houses away from the Joneses in the 1850 Vermilion County census.

Best of all for Eliza was her attentive beau, William Browne, aged twenty-three. He had left his family to sail with hers. They would marry shortly after arriving in the Grand Prairie.

Two stories remain about their journey: The five-year-old Frederick "almost fell overboard on the trip over and was saved only because he was caught by his dress tail. He wore a dress instead of a suit." according to the family story. And one of the pieces the Joneses brought with them was a large mantel mirror that was to be inherited by the oldest daughter in the family. In 1921 it was hanging in the house of Alice Elisabeth Puzey Graves.

On August 30, 1849, the Jones's friend Edwin Horniblow, whose family was mentioned in Eliza Jones's London diary and who lived in 1850 with the Thompsons, arrived alone on the *Danube*. Some Jones and Church descendants heard he hadn't liked America and returned to England.

During the next travel season, which was the spring of 1850, the pace of arrival increased. Sophia Puzey Church arrived on the *Yorktown* on May 17, 1850, with her three children, Thomas, Jane, and Sarah. Albert, a nephew of her husband, Henry, was with her. He was headed to live with his brother, George W. F. Church. On the ship were Sophia's half sister, Phillis Puzey, who was now sixteen years old, and Sophia's nephews, Albert and Henry Puzey, the twenty-two- and twenty-three-year old sons of Sophia's brother Joseph. Since Sophia's husband, Henry, had gone at least a year before and bought land, once again he was

Tossed at Sea

missing from the family group. By the 1850 census was recorded, however, Henry was living with his family.

Finally for that season of travel, on August 1, 1850, the Onleys' married son, William, landed in New York with his wife and two young children.

—〰—

When the census was taken in Vermilion County in 1850, at least 152 English-born men, women, and children were living there. A group numbering 152 out of a population of 11,504 might not seem like a lot until one compares it to Champaign County. At the time that county had 2,649 inhabitants and only nine English-born inhabitants. Isaac Sandusky's lecture would have extolled the virtues of Vermilion County's rivers, giving it an edge over Champaign County, which had fewer waterways. There was also the added advantage in Vermilion County of the presence of the English-born Manns and Corks who had immigrated to its northern part in the 1830s and early 1840s. These factors meant the Grand Prairie English set their sights on Vermilion County despite there being more land available in the sparsely settled Champaign County. That adjacent county didn't get going until the railroads came through.

It wasn't just Champaign County that attracted fewer English. Compared to most other Illinois counties, Vermilion County specialized in English immigrants. So said Douglas K. Meyer in a report entitled "Foreign Immigrants in Illinois 1850." An exception was Edwards County in southern Illinois, which was known then as Boultinghouse or English Prairie. George Flower and Morris Birkbeck settled this area just after the War of 1812. Birkbeck wrote a popular book that attracted more than two hundred other English immigrants

115

during the 1820s. Perhaps Isaac Sandusky's lecture had created enough of a stir to give Vermilion the advantage over Edwards with English immigrants in mid-century. A more likely explanation, however, was that Edwards County had less available government land.

The English who came to the Grand Prairie at the end of the 1840s and early 1850s clustered in the southern part of Vermilion County. This was northeast of where Tom and Dee Belton's farm is now and a few miles southeast of the settlement then known as Butler's Point. Isaac Sandusky lived nearby. Later they would informally call this area Fairview.

Eventually the village of Catlin subsumed the small neck of forest known as Butler's Point. The English, however, would not have identified themselves with Butler's Point but with Vermilion County, the land office at Danville, and the Grand Prairie.

They had left England halfway through the century, so they missed being a part of many of the developments currently associated with England.

They missed viewing the iconic buildings known as Parliament, which by the 1850s had been rebuilt after the fire that had consumed them two decades earlier. They missed surgeon John Snow's discovery that victims of a cholera epidemic in Soho drank from the same well on Broadwick Street. This provided evidence that cholera entered through the mouth rather than being an airborne miasma. The immigrants from Berkshire missed the birth of the lamb with two faces, which the Puzeys' relative Violet Howse described in one of her books. The Londoners missed the opening in 1863 of the world's first underground railway, which ran from Paddington to Farringdon, the first section of London's Underground. They did not know

Big Ben, which started ringing in 1859 on the towers of Parliament. They and their descendants missed England's nineteenth-century wars, the Blitz, and the hunger after World War II. Instead these families were building lives on the American prairie.

13

Claiming the Prairie

On Illinois farms today, children awake to the sounds of tractors, usually started by their fathers before they head into the fields to plant, cultivate, or harvest. The reliable tones of the engines mean that all is right with the world.

The English-born families in the Grand Prairie in the 1850s would not have heard the tractors' harmonies, but they would have listened to the songs of the meadowlarks perched on the fences, the snorting of horses or oxen, and the clanking of the plow's chains as they were attached to the animals' yokes. It must have taken several seasons before the sounds of the prairie could comfort the English immigrants, but they had to adapt quickly. The money in their pockets had to be transformed into a livelihood in America. For most of them, getting settled meant buying land.

In June 1847, shortly after he arrived, Richard Puzey bought eighty acres of land from the federal government in Township 18, Range 12, Section 14. This was four miles southeast of the small settlement at Butler's Point and a few miles northeast of Tom and Dee Belton's present-day farm. He extended his holdings with an adjacent forty-acre parcel two years later.

119

As the first English-born settler in the few square miles where the present-day townships of Catlin, Georgetown, and Carroll meet, Richard Puzey would have been the one to name the area Fairview. The name was in use by at least 1856 when Fairview School was established. Richard probably enjoyed bestowing upon his new land the name of that picturesque house in the village where he was born. In doing so, he was following a long tradition of settlers naming American sites after the places they had left behind.

The word "Fairview," however, doesn't describe what Richard Puzey first saw. His son Richard later described his father's first impressions in a newspaper interview that appeared in 1914. "Afterwards he admitted that he was never more disheartened in his life than he was upon his first ride over his newly acquired real estate," Richard Junior was quoted as saying about his father. "Although he had paid but $1.25 per acre for the land, he said time and again as he rode and walked over the barren marshes on every side that saw little to encourage agriculture, he was tempted to tear up his deed and throw it into one of the numerous sloughs which for the most part made up his land."

Richard Senior, however, might have been exaggerating for dramatic effect. He had been living with the Sanduskys for several weeks by the time he bought his land. The land he bought was near them. Government land was generally unbroken land. Surely he knew what he was getting.

Perhaps this story about Richard's dismay at the land isn't about his own plot but about seeing the Grand Prairie itself for the first time. Marshy and entangled by tall grasses, hazel bush, and blackjack oaks, it did not live up to the promises conveyed in Isaac Sandusky's lecture. The work to make it profitable would be hot, relentless, and backbreaking. There was nothing to be done, however. Richard couldn't

Claiming the Prairie

go back to England when scores of his family members and friends were planning to follow him.

The tools at his disposal weren't sophisticated. The mole plow, which bored through soil and created drainage channels, wasn't available until 1856. The cylindrical red ceramic drain tiles or polyethylene pipes that now carry away water came even later.

Richard, however, with his rural Berkshire roots, wouldn't have been without resources. English farmers were known for their skills in draining the land. The heat in which Richard had to work might have been a surprise, but he had no family to distract him. Prairie land was quicker to tame than wooded land, where it took one person more than a year to clear five acres.

Driving along the present-day country roads next to Richard Puzey's farmland and that of the other English, one can still see the deep ditches that Richard would have been the first to dig. Richard Puzey's farm ultimately was a pastoral showplace. His granddaughter Matilda's biography of him describes the two-room house he first built. It was the only frame structure for miles around. Over the years he and his son enlarged the house and tended an orchard, a walnut grove, a berry patch, asparagus and rhubarb beds, and a long row of gooseberry bushes.

The land Richard saw when he arrived was still soggy, but the Grand Prairie had become considerably more civilized during the 1840s. When Zachary Taylor took office in 1849 as the twelfth president, it was clear the United States of America had settled in for the long haul. By 1848 Illinois had come up with a new, improved state constitution, although it disappointed and wasn't really improved until 1862.

The population was exploding. Between 1840 and 1850, Illinois grew by more than 360,000 people from just under

121

Getting to Grand Prairie

a half million. Between 1850 and 1855, the population increased by more than 440,000.

English immigrants were not the most numerous in the state, but their numbers were still significant, and they were the foreigners most intent on owning land. About 13 percent of the adult males in Illinois in 1850 were foreign-born, which was slightly above the 11 percent of foreign-born immigrants in the United States as a whole. Of those 110,000 or so foreign-born adult males in Illinois in 1850, 36 percent were from Germany, 27 percent from Ireland and 18 percent from England. Many German and Irish immigrants had settled in farming communities, but they also clustered around the well-established St. Louis and the growing city of Chicago, seeking nonagricultural jobs. While English immigrants certainly populated Cook County and the area around St. Louis, more of them preferred rural to city life.

Other counties in northern, southern, and western Illinois had successfully attracted earlier English settlers intent on farming. However, the Grand Prairie, the remnant of the ancient glacial lake bed located mostly in east-central Illinois, had fewer navigable rivers than other parts of Illinois, and it had that marshy problem that postponed its appeal. It wasn't intensely settled until government land was running out in other parts of the state. The English immigrants already in residence in the northern part of the county and the shallow Vermilion River, which nevertheless could carry livestock and produce to the Wabash where they could go upriver to the market town of Covington, Indiana, offered consolation. Therefore, despite Richard Puzey's complaints, he and the other English arrivals around 1850 and after would make do with the undrained land still available in Vermilion County.

122

Claiming the Prairie

Native-born Americans, of course, outnumbered immigrants. Of the 11,504 residents counted in the 1850 Vermilion County census, 4,971 residents were born in Illinois. Almost all the native-born, however, were the young children of parents born elsewhere. After Illinois, Ohio contributed the largest number of settlers with 1,838 coming from that state. Kentucky was next with 925 Illinois residents born there. Indiana-born residents numbered 791. Virginia contributed 602 residents and North Carolina 154, but few came from other Deep South states. Seventy-one immigrants came from other European countries, Australia, and Canada, but the 152 English immigrants far outnumbered other foreigners. More would arrive after 1850. "No other state, not even New York, experienced so large a net increase in its English-born population as did Illinois," said historian Charlotte Erickson about the 1850s in *Invisible Immigrants.*

The population increases reflected the development of what had been the frontier into a more structured society and, in turn, caused more development. The influx of widely traveled English immigrants lent new ideas and contributed to the Grand Prairie's advantages and growing sophistication.

In *Britain to America*, William Van Vugt quotes from an 1840s letter written by an English-born woman in Monroe County, Illinois, to her family back in England. She described her farm:

Our house stands at the top of a hill and we have bought eighty acres of land. We have got ten acres of wheat which will be ready in August, and we are getting ready for our garden and potatoes and Indian corn. We have two cows, four calves, two horses,

123

twenty-four pigs, two dogs and I cannot tell you the number of chickens. Before you get this letter we shall have more young pigs. So you see our stock keeps increasing continually. We killed a fine fat pig for Christmas. We had some mince pies. I wished many a time we could send you some.

The dependable agricultural bounty gradually fostered the development of commerce and cultural, religious, and educational institutions and transformed the former frontier. Vermilion County's Presbyterian and Methodist churches were still meeting in improvised schoolhouses, but that began to change in the 1850s.

The salt works, an attraction of early settlement, had largely shut down by the end of the 1840s. Better salt fields had been found in western Illinois. Danville's federal land office, however, took the sting out of losing the salt industry. The land office, plus Danville's eventual role as the county seat, attracted clerks and lawyers. Among them was Abraham Lincoln.

New settlers were arriving every day. A lot of business had to be done buying and selling household goods, farm machinery, wagons, and carriages. Problems had to be solved to successfully transport wheat, apples, and livestock to the rapidly growing city of Chicago and the older cities of St. Louis and New Orleans. Agricultural fairs sprang up—one near Butler's Point. By 1850 Benjamin Canaday, an early settler from North Carolina, had erected a large three-story store at a cost of five thousand dollars in Georgetown in the southern part of Vermilion County, showing the confidence Canaday had in the future prosperity of the region.

Claiming the Prairie

Like Richard Puzey, the other English-born settlers bought land within days of their arrival. Thomas Onley bought 160 acres of federal land on April 27, 1849, in the section next to Richard Puzey. Onley and his family had arrived in New York only three weeks earlier. They seemed to have mixed feelings about the Grand Prairie, though. They and other family members appear in later 1800s census records in Indiana as much as in Illinois.

Henry Church bought 160 acres of federal land next to Richard Puzey on May 8, 1949, a full year before his wife, Sophia Puzey, and her entourage arrived in New York Harbor. Sophia must have been or grown into an authoritative, independent woman, since her husband was rarely to be found.

The English and other immigrants arriving from Indiana, Ohio, New England, and Kentucky caused sales of federal land in present-day Catlin Township to reach eighty-one parcels containing more than 5,100 acres from 1848 to 1851. Some of the acreage was bought by means of warrants from Revolutionary War, War of 1812, or Blackhawk War veterans, their widows, or speculators who bought such warrants at cut-rate prices from veterans with no intentions of leaving their East Coast homes for the Northwest Territory. The warrants had been issued to these people as rewards for their service and could be traded for government land.

—ᚳᚳ—

On September 6, 1849, Thomas Bentley bought his first parcel, forty acres northeast of Butler's Point. This was a couple miles away from the Puzeys and the Churches, but it was near where the Hinds settled and where Henry Jones

Getting to Grand Prairie

would later buy land that his daughter Eliza lived on. Five days later Bentley bought another eighty acres.

Bentley bought his land for four dollars an acre with another man, William A. Rawlins, whose story isn't known. The seller was Isaac Sandusky. The lecture Isaac had paid for in London promoting Vermilion County was paying off for him. Between 1830 and 1851, Sandusky had bought at least 1,227 acres of land in Townships 18 North and 19 North and Range 12 West. The sale to Thomas Bentley in 1849 was only the first of many he made to the people he had invited to his part of the Grand Prairie.

The higher price Bentley paid was typical for land in private rather than federal hands. For one thing, the seller might have already broken the land, built a house or a barn, or otherwise improved it. There also seemed to be a tacit agreement among the buyers that a private seller was entitled to a price higher than the federal government.

Since prairie land had finally been recognized as fertile, by the 1850s it was more expensive than timberland. Many shared the dream that the price of land would increase, and sellers would eventually make killings on their investments. Easterners were particularly susceptible to this dream. The trustees of Dartmouth College in New Hampshire invested early in about 340 acres of federal land in Vermilion County. They sold the parcel to Richard Foxe, a resident of Pennsylvania, in 1853 for $1,920. That was $5.65 per acre. He too must have bought it as an investment. There is no record of him living on it or farming it.

The English would have had to learn the uniquely American system by which the land in the Northwest Territory was measured. Working from baselines and principal meridians, surveyors divided the land consecutively into townships. Each was six by six miles square. Then they

126

Claiming the Prairie

divided the townships into thirty-six numbered sections of one square mile (640 acres) each. A section could be divided further into halves, quarters, and smaller fractions.

To find the exact location of a piece of land, one had to first find its township. That was the equivalent of its latitude, counting from a baseline running east and west. The land settled by the English families coming to the Grand Prairie in this period was eighteen and nineteen townships north of the baseline in southern Illinois.

That's only half the story, though. One must also locate its range, or the equivalent of its longitude, by counting from the nearest principal meridian running north and south. In eastern Illinois that north-south line is in Indiana. When counting from the principal meridian, one must still count townships, but when counting east and west, townships are called ranges. The English immigrants bought land in the 1840s and 1850s primarily in Range 12 West. That was twelve ranges (townships) west of the second principal meridian.

Once the township and range had been pinpointed, the next step was to locate the number of the section. Numbering began in the northeast corner of the township and alternated direction in each row until section thirty-six reached the southeast corner. A deed specified how much of the section and in what part of the section the land lay.

Complicating matters was that the named administrative townships—Catlin, for example—do not correspond to the numbered land measurement called a township. Catlin Township, therefore, contains both the northern part of Township 18 North and the southern part of Township 19 North within its boundaries. In addition, Catlin Township covers the western part of Range 12 West and the eastern part of Range 13 West.

127

Getting to Grand Prairie

At the time most of the Grand Prairie English bought their land, Vermilion County was divided into only two parts. Ripley was to the north, and Carroll was to the south. In November 1851, the county was divided into nine townships—Danville, Georgetown, Elwood, Carroll, Vance, Pilot, Middlefork, Ross, and Newell.

The English who immigrated around 1850 had mostly settled in what was the confluence of Danville, Georgetown, and Carroll. At that time Danville Township contained the area later established as Catlin Township. Immigrants would certainly have known of the town of Danville, since the federal land office was there. They would have known they were in Vermilion County. They could have identified Butler's Point to the northwest and Brooks Point to the northeast. Mostly, though, they still recorded their location as the Grand Prairie.

At the same time Thomas Bentley began buying land, Henry Jones did too. On September 10, 1849, Henry bought eight hundred acres containing an 1826 brick house from an old settler, Francis Whitcomb. Henry is said to have added two large rooms to the house. The house still stands across from the Catlin Historical Society on the old Fort Clark Road, which ran east and west between the Indiana state line and Fort Clark on the Illinois River at present-day Peoria. Whitcomb financed the purchase himself with instructions that Henry pay him by January 10, 1850, or he would take the land back. Whitcomb stipulated that the Methodist meeting house, located on one-quarter of an acre in that area, was not part of the sale. Henry surely was eager to get into a house of his own. His large family, after

Claiming the Prairie

all, had arrived a few weeks before. Those "trustees" whom he wrote about in his shipboard letter to his family must have sent him money before the January deadline, because he was able to pay Whitcomb on time.

In 1850 Henry stepped up his pace of acquiring land. He bought eighty acres from Noah Guymon at four dollars and fifty cents an acre and acquired a "ditching machine," presumably to assist in drainage, from a deceased farmer's estate. In May he bought 480 acres of government land in separate lots west of Richard Puzey's and Thomas Onley's land. Henry must have sold land too. The 1860 U.S. agricultural census shows Henry Jones owned 420 acres of improved land and eighty acres of unimproved land worth $25,100 in Catlin Township alone. By the end of his life, according to the family story, Henry owned more than three thousand acres, fourteen head of oxen and a considerable number of cattle. Some records of Henry's acquisitions name only "parts" of a section rather than the exact acreage or location. However, the specific records of land transactions show Henry as close to that total by the time he died. Land records show he spent at least $24,572 on farmland and house lots in what became Catlin Township and the village of Catlin over twelve years. He might, however, not have owned all those three thousand acres at the same time.

A national currency with the same value throughout the country had not taken hold yet in the United States. Henry and the others, therefore, would have at least initially purchased the land in gold. Historian Charlotte Erickson quoted one letter writer emphasizing this in a letter back home: "[T]he only money in the country seems to be brought out by Englishmen."

Getting to Grand Prairie

By 1852 some of the younger English immigrants were acquiring land. The twenty-four-year-old Albert Puzey bought eighty acres of government land in June, shortly before he returned to England to find a bride. George W. F. Church, aged twenty-two, bought an unspecified amount of land in May 1852 for $110, and Albert Church, George W. F.'s nineteen-year-old brother, paid $1,600 for several parcels of land formerly owned by a family named Songer. The amount of money the young Church men were spending showed the significant sums their family in England had sent them for buying land on the Grand Prairie. They had to have cash for land purchases. Mortgages or financing options such as Francis Whitcomb provided for Henry Jones were uncommon.

It was a good thing Henry Jones had some land. Six days after he bought Francis Whitcomb's house and adjacent land, his wife's mother, Ann Hough, died, and he needed a proper burial ground. He set aside an acre or so across the road from his house, and Ann was the first person to be buried in what became known as Jones Grove Cemetery or sometimes the English Cemetery.

The second person to need a burial was the Thompsons' child Ann. A local newspaper article written in the later nineteenth century described what happened:

One day she was drying some doll clothes before an open fire in the grate when her clothing caught fire, and she was so badly burned that death ensued within a few hours. Frederick Jones, who resides here, was at the front door of the Thompson home when the accident happened and well remembers seeing the frantic mother throwing the quilt over the head of the child in her efforts to extinguish the flames. She finally succeeded but the little sufferer passed away in terrible agony within a short time.

Even today descendants of the English settlers of the 1840s and 1850s are laid to rest in this peaceful spot between the timberlands and the prairie.

—∞—

While the English families were busy buying land, their adult children were busy marrying. They didn't stray far from their own kind. For two decades most of the young people married other English immigrants and ignored their American-born neighbors.

The fifty-one-year-old Richard Puzey married on January 14, 1849. This was before most of his friends and relatives arrived and almost two years after he came to America. He chose thirty-year-old Amelia Jane Eustice Cork, another English immigrant and a sister of the Reverend William Cork of Rossville. The Corks, whose later family members tended to call themselves "Cook," were neighbors of Abraham Mann in Ross Township.

Amelia's home in Rossville, in the northern part of Vermilion County, must have been a good day's ride from Richard's place. The physical distance between the two was apparently less than the cultural distance the newlyweds felt toward eligible American-born neighbors.

A particularly happy occasion was when Eliza Jones married William Browne, the attentive beau who had followed her across the Atlantic. Although Eliza had had many adventures since the last entry in her diary in early 1849, she did not inscribe any additional notes in it until she marked this occasion on the inside of the back cover: "William Browne and Eliza Jones. Married May 16, 1850." Her sister Sarah married George W. F. Church in November of that year.

The habit of the English preferring other English immigrants for marriage would continue with only a few deviations until after the Civil War.

14

The Special Case of Will Clipson

The Clipsons must have missed the Bentleys, Puzeys, Churches, and especially the Joneses as those families left London for the Grand Prairie. By late 1850 all of them were gone, but the Clipsons still showed no signs of leaving England. They waited until events forced them out.

In 1851 the census taker recorded William and Matilda A. with their surviving children, Catherine, Jane, William, John, and James. A servant, Elizabeth Truscott, aged twenty-two, from Cornwall lived with them. The census revealed Will Clipson's changed occupation. He reported himself as a retired licensed victualler. He had shed the job of publican so he could concentrate on running betting games.

The census located the Clipsons at the end of March at Richmond Terrace in Southwark. By July 15, though, when their daughter Harriett Ann was born, they were living and Will was working on the north side of the Thames at 23 King Street, St. Paul's Covent Garden. This was Henry Jones's old address, sometimes listed as the adjacent Rose Street, for his gas fitting business.

Betting games, then as now, sometimes attracted a criminal element. One of Will's employees later figured in

133

a famous 1855 train robbery. The event was chronicled by a man named Samuel Smiles who told the story of William Pearse, later spelled "Pierce." "I met Pearse by accident just by Covent Garden Market," he wrote. "He was then a clerk in the betting office of Clipson, King Street. I went there occasionally and made several bets."

Pearse had bigger things in mind than earning a living as a lowly clerk. He staked out the gold shipments on the South-East Railway Company, acquired a railway employee as an accomplice, copied keys, hopped on the train, and opened the trunks carrying the gold. After he transferred the gold into his bags, he filled the trunks with lead shot and relocked them. The theft was not discovered until the heavy trunks were opened in France.

Smiles was only the first of many authors to tell Pearse's story. Michael Crichton's *The Great Train Robbery* was made into a movie in which Sean Connery played the part of Will Clipson's clever but crooked former employee. Will would have missed the robbery and the excitement over it because he had emigrated by then.

Smiles, a gentleman, openly admitted he was a gambler. Betting was, after all, a time-honored London tradition that the "better" classes and lesser mortals enjoyed. The higher classes, however, often kept to their clubs for betting, since they didn't want to mix with their social inferiors. It was natural to pair betting with horse racing, a sport popular in England since the twelfth century.

Not that reformers didn't frown on betting. Then as now they maintained that betting houses tempted those who could ill afford to lose and also encouraged fraud and dishonesty.

Will objected strongly to such interference. In the early 1840s, he had joined Tattersall's, the auction house

The Special Case of Will Clipson

for horses near Hyde Park Corner. Connected to it was the Turf Subscription Betting Room, which was the authority on the rules of betting on the turf. Will was well-known as a bookmaker and continued to place ads for the games he ran through Tattersall's.

On July 6, 1852, he and thirteen other betting office proprietors sent a letter to Her Majesty's First Lord of the Treasury, the Earl of Derby, the man for whom Derby Day was named. The letter was published in *The Racing Times* and other newspapers. The proprietors said they had been "the subjects of prejudice and ignorance" and that "they are prepared to prove that their transactions have been conducted upon those fair and honourable principles which all true lovers of the great national sport of horse-racing must wish to see in turf matters." They regretted "the efforts of misdirected zeal to suppress their establishments."

Despite their efforts Parliament suppressed the betting houses in the Betting Act of 1853. It didn't matter to Will. By then he had made his way to America. Perhaps he would have left England anyway, since the suppression of the betting shops would have restricted his lucrative occupation. In October, however, he had to leave London. He was run out of town.

The flurry of newspaper accounts started October 18, 1852. They accused Will and five other betting house proprietors, some of whom had signed the same letter he did, of cheating winners who had placed bets on a race at the Newmarket Cesarewitch. Because so many betting houses were involved, it is possible that as a group the proprietors hadn't been paid by Tattersall's or with whomever they had placed their clients' money. No trial was ever held, so the facts surrounding the accusation have not seen the light of day. But the betting house customers were furious. The day after the first accounts appeared, the newspapers reported that Will and the other betting office owners had fled. Some were accused of disguising themselves. Some of those who said they were swindled recommended the perpetrators be horsewhipped. It's not obvious from the newspaper reports how much money Will and the others were accused of making away with, but one report said it was around £20,000.

No matter. The Clipsons were gone from London. They fetched up near the home of Matilda's sister Emma and her husband, William Dickinson, in Langrickville, a village near Boston in Will's old county of Lincolnshire. Will and Matilda's son, a second Richard, was born there in December.

Will's sister Rebecca was living nearby in East Keal where the fens and wolds meet. She had married John Johnson but had no children. His sister Sarah's whereabouts are unknown. John Clipson, Will's father, had died in October 1850, and his mother, Charlotte, was living with Rebecca and John. Will's sister Lucy had married John Bogg and was living in nearby Raithby with him and their three children.

Having family nearby could have been comforting, but Will feared he would be found. By late March he, Matilda, and their family, along with Emma and William Dickinson and their family, journeyed to Liverpool, boarded the first-class section of the *Siddons* at Waterloo Dock, and were on their way to America. Will wasn't about to board a ship leaving from London, where disgruntled victims lay in wait for him. After landing in New York, Will may have telegraphed Henry Jones to let him know he would soon be arriving in Illinois, since Danville had a telegraph office by 1852.

Emma and Matilda left their mother and father, who were still living near Boston. Will left his mother and at least two sisters. They would not see one another again.

The Special Case of Will Clipson

On that trip the *Siddons* carried 422 passengers. Half of them were Irish, many alone and in steerage. A quarter of the steerage passengers were German, representing the continuing great influx of immigrants from that country as the century wore on. The English passengers mostly came with families of six to eight children. A few Scots and one passenger from Holland rounded out the roster.

The Clipson family story claims that Matilda had her hands full with Will and baby Richard both sick. "This baby son had been very ill aboard ship, and the captain had told his mother that in case of death that the child would have to be buried at sea," according to the family story. "She objected and vowed that she would jump overboard. So, in appeasement, the captain told the ship's carpenter to make a casket and he promised that in case little Richard did not live, that he would allow him to be brought into port. But the baby lived."

It's hard to imagine making room on a ship for the number of household items they took with them. And they couldn't have been the only immigrants arriving with such a load. In 1889 Wilse Tilton was a social reporter and writer of the column "Catlin Clack" in the *Danville Weekly News.* He supplied a partial inventory of these items when he covered the reception for about two hundred guests that Matilda hosted for her youngest, American-born son on his wedding day.

"The Clipson mansion contains a great many curiosities and keepsakes of English origin," he wrote, "among which is a bronze statue presented to the father by the London Gas Company. He was then its president." This title is quite different from the one Will provided to the census takers.

Tilton described the two large oil paintings of Matilda and Will that they sat for shortly after their marriage. One descendant said the Clipsons carried with them cases of wine and other drinks. These items would be helpful if Will were to establish a pub in his new country. Other descendants claimed to possess the trick candlestick holder that Will was supposed to have given Matilda upon the birth of their first son, as well as a stand with a hidden compartment. Harriett Ann's will listed a large antique mirror and silverware that her parents had brought from England.

When they arrived in New York, the Clipsons and Dickinsons were the first to disembark. Will identified his occupation as an engineer, possibly to confuse anyone who might follow him to America.

When the fifteen family members got off the ship, they might have heard two pieces of news. Shortly after they left, the steeple of the East Keal church had fallen "down with a tremendous crash, forcing the earth up into an embankment to the height of three or four feet," according to a local account. This village was the home of Will's sister Rebecca and their mother, Charlotte. Having been left behind by their brother and son, did Rebecca and Charlotte view it as a metaphor for the collapse of their world? If so, Charlotte didn't have long to contemplate it. Two days before the *Siddons* landed in New York, she died. Her death was the last mention of a Clipson in Miningsby church records.

—⁓—

Will, Matilda, and their children did not tell their descendants why they had had to leave England so abruptly. It is not known how much their friends in America knew about Will's problems in London. The arrival in the Grand Prairie of the Clipsons and Dickinsons, however, must have been big news for the growing English community there. The two new families fit right in with the friends they hadn't seen for three or more years.

Like the earlier arrivals, the Clipsons didn't wait long to buy land. In their case, however, it might have been with ill-gotten gain. On July 2, 1853, Will bought one hundred and sixty acres between Richard Puzey's farm and land Henry Jones owned from Thomas and Rachel Onley, who had bought the land from the federal government in 1849. Most likely the Onleys had paid $1.25 an acre, which was the going rate for federal land at the time. They sold it to Will for $1,100 or $6.875 an acre—five and a half times the amount they had paid.

According to the 1889 *Portrait and Biographical Album, Vermilion County, Illinois,* the Clipsons began farming and immediately built a house. Carpenters and builders had been plentiful since Vermilion County's early days, although their skills were uneven. Timber was readily available, and several sawmills were in operation. As early as the 1830s, Dr. William Fithian, Abraham Lincoln's Danville friend, had built a house with planed floors rather than puncheon, the name for rough timber. By the 1840s, settlers were building balloon-framed houses rather than the log cabins they had previously erected.

On August 2 Will bought forty more acres for four dollars an acre just north of his purchase a month earlier. This time the sellers were Henry and Marietta Ellsworth, land speculators from Connecticut who also had a residence in Lafayette, Indiana. Ellsworth, a lawyer with an interest in improving agriculture, had served as the first commissioner of the U.S. Patent Office. The Ellsworths did all right in their transaction with Will. They had bought this parcel

Getting to Grand Prairie

only a few days before from another Connecticut land speculator, David Watkinson, for one dollar an acre, making a profit of 400 percent.

Was Will the innocent newcomer the more experienced locals had tricked? It cannot be known for sure, but he must have had advice from Henry Jones about his purchases. Henry might have even lined them up for him when he got word of Will's impending arrival. At the price the Clipsons paid, the land must have been partially broken, and maybe there was a farm building or two. The out-of-state speculators might have been happy to turn a sale around fast. Absentee owners often found it challenging to manage their investments. Squatters sometimes settled on land, breaking it, and farming it without the owner's knowledge. Removing a squatter was hard to do from far away.

Another explanation is that those involved believed land would always go up in value in such a flourishing place as the Grand Prairie. Therefore, a high price didn't matter in the long run. In any case, in December 1857, Will and Matilda sold this last parcel to Henry Jones for $1,400, or thirty-five dollars an acre—not a bad return for holding it for only four years.

—⁓—

Rachel Onley, Marietta Ellsworth, and Matilda Clipson each had her name on the deed when the land was sold. Both the husband and wife typically owned the land in Illinois. If the husband owned solely, at his death the wife usually would receive income through a "life estate" in the land before it passed to their children upon her death. Daughters and sons would typically inherit real estate equally. This was unlike in England where the oldest son was the heir. The English land inheritance system had also functioned in Virginia and other longer-settled American regions, but it was dying out. The Grand Prairie English women had typically lost servants in coming to America, but they had gained land ownership. One wonders if they thought it was a fair trade.

—⚹—

Will Clipson wasn't the only one buying land in 1853. That year Vermilion County recorded 965 deeds. The Catlin Historical Society's records show that seven parcels changed hands in what became Catlin Township. The total acreage was 330.68. Naturally Henry Jones accounted for a portion of that. He had bought eighty more acres for $320 on June 18 from David Watkinson's brother, Robert, and his wife, Maria, of Hartford, Connecticut. Henry now had at least 560 acres.

About a year later, the Clipsons were settled. On November 26, 1854, Albert (Tal) was born. He was their only surviving American-born child. They recorded his birth and their address in their Bible. The place of birth was Globe Farm, Grand Prairie, Illinois. They had named the farm after Will's favorite pub. Apparently they didn't deem it important to mention the county or township in which they were living.

After Will Clipson's arrival, he found himself threatened. Roy Clipson, the youngest son of Will and Matilda's son James, recounted the matter in a 1964 letter to one of the Clipson family history writers. She conveyed the essence of Roy's letter:

The farm [the Clipsons] had settled upon was approximately three and a half miles south of Catlin.

Getting to Grand Prairie

About three-quarters of a mile northeast there was at the time a walnut grove. Soon after William Henry had come to this country, a Frenchman came to this country looking for him. He came out to the farm and challenged William Henry to a duel, and our ancestor chose swords. They chose their seconds, had a doctor near, and at daybreak the match was begun. The Frenchman was wounded in the arm and returned to France. What it was to settle no one knows, as William Henry did not even mention it to his wife.

Roy admitted he could not verify the story. Another descendant claimed the weapons were pistols.

Tales such as these are suspect, but this one has the ring of truth. Puzey family records speak of their walnut grove, and the Puzeys lived next door to the Clipsons. Furthermore, no matter what the weapons were, the word "Frenchman" added credibility. In the first half of the 1800s, historian Jerry White reports that Londoners called every foreigner a "Frenchman" no matter where he came from. Had this foreigner or "Frenchman" been cheated, or thought he had been cheated, by Will Clipson? Had he followed him to Illinois? Clipson family historians claim that Will was always concerned about what they sometimes termed the "British mafia," which he thought was out to get him. It is doubtful the real story will ever emerge, but none of Will's descendants would have naturally used the word "Frenchman" in that way unless Will himself had used it in the telling.

Matilda Clipson's sister Emma and her husband, William Dickinson, lived on a rented farm during those early years. Evidently they lacked the cash to buy land

The Special Case of Will Clipson

when they first arrived. The 1889 *Portrait and Biographical Album, Vermilion County, Illinois* places the Dickinsons in Township 19, Range 12, Section 26. This was northeast of Henry Jones's house in the village. They bought that parcel and parts of sections twenty-seven and thirty-four for ten thousand dollars in 1875. The recorded deed does not give the exact acreage, but the county biography said the Dickinsons' farm totaled 197 acres. That would have made the selling price about fifty dollars an acre. In the same year the Vermilion County records show that Richard Puzey paid thirty-three dollars an acre for 120 acres he purchased. Although it took some time to amass enough cash to buy their land, the Dickinsons had clearly prospered. Land values had increased. The Dickinsons' land was more valuable because of the sweat they had put into it when they didn't own it, but that was part of the deal.

15

Ambivalent Citizens

Despite their haste to buy land, the Grand Prairie English continued to be of two minds about America. It might be a fine place to establish a farm, but would you really want to marry an American? Maybe the best idea was to go back home and find a young English woman who was looking for adventure.

At some point after he bought his eighty acres, Albert Puzey did just that. He returned to England for what must have been a bittersweet reunion with his parents, Joseph and Beatrice, since soon he would be leaving again. In April 1853, he married Eliza Dowding. According to the family story, she was a concert pianist. One wonders, though, why a concert pianist would agree to move to a recently settled frontier with few opportunities to perform. Albert met her through his mother's Blanche family relatives who lived near her in Gloucestershire.

Shortly after their marriage, Albert and Eliza embarked from Bristol for New York where they landed on July 13, 1853. Accompanying them was Albert's relative, Hester Blanche, aged thirty.

Getting to Grand Prairie

A young man named Alfred Church, aged eighteen, was also on the ship's manifest. It is possible this is really Adolphus Church. There would have been nobody better for Adolphus to travel with to America than a Puzey who was his cousin by marriage and the nephew of Uncle Henry Church's wife, Sophia Puzey. Adolphus later verified that he immigrated in 1853, but no Adolphus appears in ship records. The immigrant ships' captains, who usually prepared the manifests, often got names wrong. Adolphus was certainly in Vermilion County by 1854 because his father addressed a letter to him in March of that year. The letter described his plans for a fishpond and fountain in his London garden and asked how Adolphus's money was holding out.

Also immigrating that summer was a twenty-nine-year-old cousin of the Puzeys, Frederick Tarrant. He was born in their hometown of Stanford in the Vale in Berkshire.

Fred's journal shows he left Bristol, England, in steerage on the *Cosmo* in early April with fifteen other passengers whom he knew from Berkshire. The night before he boarded the ship, he slept in the same room with nine of his friends at a local pub. They took with them mutton, cooked and salted beef, bacon, and other types of pork, cheese, butter, beer, potatoes, plates, cutlery, commodes, mops, kettles, and tobacco. The ship provided water, tea, sugar, rice, and flour. During the voyage a child was born to one of the passengers and named Cosmo after the ship.

While they were cooking their first dinner on the ship before it left, Edwin Puzey, Henry and Albert's younger brother, visited them. He was on his way from Upper Mill, the location of Joseph and Beatrice Puzey's brick kiln. He intended to visit his mother's brother in Latteridge, Gloucestershire.

146

Ambivalent Citizens

Fred's ship, alternately buffeted by severe weather and subjected to dead calm, took seven weeks to make the journey to New York. After he landed, Fred first went to Detroit where his Berkshire shipmates were headed. He arrived in Vermilion County in early summer.

—ᴍ—

Illinois had plenty of fresh air, clean water, and a surfeit of land. The Grand Prairie, however, did not spare the immigrants from diseases and disasters. Grandma Lura Guymon practiced the healing arts and had a reputation for the safe delivery of babies, but few doctors were available. The Grand Prairie English, and everyone else there, were on their own.

In 1852 one-year-old Eliza Browne, Eliza Jones and William Browne's first child, joined Ann Hough and little Ann Thompson, who had burned to death, in the English cemetery. It wasn't only Eliza Jones's child who suffered misfortune. On March 24, 1854, the beloved William Browne, now the father of one-year-old Emily, also called Emma, was unloading hay from a wagon near the residence of his father-in-law, Henry Jones. According to the family story, "He was on the ground at the rear of the load of hay, [loosening] the pole that bound the hay. The horses became frightened and started to run away. While he was trying to catch them he was struck across the heart by the ladder of the hay, and in less than an hour he was dead."

This tragedy must have devastated Eliza, but she had a child to provide for, and it was lonely at night on the prairie. According to the family story, she married again within two months of the tragedy to the newly arrived Frederick Tarrant.

147

Getting to Grand Prairie

It was not unheard of for women to marry again within a few weeks of a husband's death. In this case, even with a relatively well-off father, Eliza and her daughter needed financial support and companionship. Men such as Frederick, who turned out to be a good provider, could be more successful with wives and children helping them with their work. The few records that exist showing Eliza's subsequent life, however, imply a kind of restlessness. Fred and Eliza moved several times and changed businesses. Eliza speculated in land well into the late 1800s. Was that restlessness a cause of her quick remarriage, a result of beloved William's death, or nostalgia for her busy life in London? It might have been all of the above.

—⚏—

The English continued marrying other English immigrants throughout the 1850s. In June 1853, Phillis Puzey married English-born Charles Holmes. He was not, however, part of the London group. If her half sister, Sophia Puzey Church, treated her like a servant, as the family story goes, she must have been happy to get out of that house. She lived a long life and had borne ten surviving children by 1882.

In December 1854, Emily Jones married Albert Church. In August 1855, Sarah Bentley married Richard Jones. Jane Clipson married John Swannell in May 1857. John's first wife, Ohio-born Ellen Elbertson, had died, and he had a young daughter, Eva, who needed care. Seventeen-year-old Alice Bentley married newly arrived John Todd, William Dickinson's boyhood friend from Lincolnshire, in 1859.

Another English marriage in 1857 was that of Adolphus Bellingham Church and Fanny Bentley, Alice's older half sister. Richard and Sarah Bentley Jones hosted the wedding

148

Ambivalent Citizens

at their house. Richard, Sarah, and Thomas Bentley, Fanny and Sarah's brother, served as witnesses.

Adolphus was called Dolf (variously spelled Doff or Doffy) by relatives back in England but became known as A. B. in Illinois. He seems to have had a romantic touch. The Church family papers include a poem A. B. presented to Fanny as a marriage proposal. A. B.'s poem goes like this:

> I'd live a life among the hills.
> Come, Fanny, wilt thou live with me?
> We'll have the music of the rills
> Or skylark's sweeter melody.
>
> No angry words shall mar our rest,
> T'were hard if two could not agree,
> We'll be so happy in our nest.
> Come, Fanny, wilt thou live with me?
>
> I know a dear sequestered nook
> A sheltered spot, a happy place.
> Oh, there the very flowers would look
> More lovely, gazing on thy face.
>
> I know a cottage far away
> Best not too far for Love to flee
> There, Fanny, answer yea or nay.
> Sweet Fanny, wilt thou live with me?
>
> The town is not a place for rest,
> I'm weary of its garish strife
> And long with nature to be blest
> And thee, dear Fanny, as a wife.

How gladly through life's checkered day
I'd share its weak and woe with thee.
Then, Fanny, answer yea or nay.
Sweet Fanny, wilt thou live with me?

It would have been unlikely for a man on the low, rolling Grand Prairie to ask his beloved to "live a life among the hills." A. B. found this sweet plea to a lass named Mary in a book by the now obscure eighteenth-century poet John Dennis, and he substituted Fanny's name for Mary's. It was a romantic beginning to what appears to have been the long, happy marriage A. B. and Fanny enjoyed.

—∞—

Besides John Swannell and his first wife, only three other English immigrants married American-born neighbors in the 1850s. Carrie Clipson married William Moore in 1856. Rachel Onley married John Cannon in 1857. Jane Church married Francis Champion in 1859. *The Era,* the London newspaper in which Will Clipson had placed so many advertisements for betting games, carried the announcement of Carrie's wedding: "15th of June. Marriage at residence of William Clipson, Grand Prairie, Vermillion [*sic*] County, Illinois by the Rev. Mr. Long of Homer. Mr. William Mellican Moore, merchant, of Georgetown, to Catherine, eldest daughter."

Evidently Will was feeling safe enough to reveal his location, or perhaps he was so homesick for his former friends he had to let them know about the changes in the family's life in this official way.

An English-born American resident wasn't enough for Henry Puzey. He returned to England, possibly

Ambivalent Citizens

permanently, since the Vermilion County biography reports Henry felt he had made a mistake in coming to America—a claim the usually positive biography writers would have had a hard time admitting. His sister Ann, who had remained in England, had married Charles Rymer. Ann introduced Henry to Charles's sister, Hannah.

Henry and Hannah Rymer were married on March 18, 1858, at her home at Wibden Farm in the parish of Tidenham in Gloucestershire. Perhaps Hannah was seeking adventure, so Henry reconsidered his attitude toward America. Perhaps upon reacquainting himself with England, Henry came to appreciate the Grand Prairie's offerings. In any case, Henry and Hannah crossed the Atlantic together later that year. They arrived at Castle Garden, the rudimentary immigrant-processing center on the tip of Manhattan that was a precursor to Ellis Island. Dealing with immigrants had become more official in the decade since the first Grand Prairie English made their arrivals.

English census records from 1851 show that Hannah's father had about 550 acres of land and three servants. His relative wealth provides evidence for the story that Hannah brought her own money to the Grand Prairie. By the 1860 census, Henry and Hannah were sitting on two thousand dollars in real estate and seven hundred dollars in personal property. Henry Puzey would eventually become one of the county's most successful farmers. He was among the first to install the new clay tiles, and he eventually possessed more than a thousand acres and a reputation as a skillful breeder of Berkshire hogs and Cotswold and Shropshire sheep. He had come thousands of miles but had not strayed far from his English counties in his farm practices.

Getting to Grand Prairie

The growing conflict between the North and the South would have soon become apparent to the English. English men and women were antislavery. It was one reason the English chose Illinois rather than a slave state or a territory in which the practice of slavery had yet to be determined. The tension between slaveholding states and free states, however, must have made the English question their decision to immigrate to America.

As the English were headed to the Grand Prairie in significant numbers, Illinois Senator Stephen Douglas helped pass a piece of legislation called the Compromise of 1850. This admitted California to the United States as a free state, allowed the territories of Utah and New Mexico to choose whether they would be slave or free, established a harsh fugitive slave law, and made various other provisions. That Douglas embraced such a muddled law might have been a reflection of the contradictory Illinois electorate of the time.

Southern Illinois residents tended to be proslavery, while folks in northern Illinois were unabashedly abolitionist. The Grand Prairie English had landed in the middle of the two factions.

The 1850 census showed that 3,773 of Vermilion County residents were born in the North. Southern-born residents totaled 798, and 1,391 came from border states. The foreign-born population numbered 223. The rest, most of whom were children of parents from elsewhere, were born in Illinois or their birthplaces could not be determined.

The preponderance of Northerners in the county might have meant that sympathies were with the North in the coming conflict. A stop on the Underground Railroad was even rumored to exist in the southern part of the county.

Although the number of American-born residents from border and southern states was substantial, their

Ambivalent Citizens

sentiments could have been complicated. Their move to a slave-free state might have indicated they weren't in agreement with those back home. Even if they were not opposed to slavery, some of those born in the border states or what became the Confederacy might have foreseen that war was probable. Selling out and buying better land in a less-troubled location would have been prudent.

Thomas Alfred and Ivea Allen Taylor headed one such American-born family. They moved to Vermilion County in 1853, a year that saw a number of new English settlers too. Thomas was born in Ohio County, Kentucky. He became a tanner and moved to Indiana. There he married Ivea, also Kentucky-born. The couple had several children before moving to Illinois and buying farmland.

The Taylors prospered in Illinois in a way that was difficult to do in Kentucky at the time. To be sure, there were vast plantations south of the Ohio River, but the majority of the white population did not live in luxury. Alexis de Tocqueville was only one of many authors who commented on the stark contrast between the tidy, prosperous farms on the north bank of the Ohio River and the scruffy establishments on the south. Tocqueville said a person on the south side who is "eager or instructed...either does nothing or crosses over into Ohio so that he can profit by his industry, and do without shame."

By 1860 Thomas Taylor was fifty-four, and he had certainly profited. He had amassed $23,000 worth of Illinois prairie and $4,000 worth of personal property, making him the fifth richest man in the section of Vermilion County now known as Catlin Township. He had done this without the help of slaves but with his family members and several hired farmhands, one of whom was a free, twenty-one-year-old illiterate African American man named James Blue.

Getting to Grand Prairie

The acquisitive Henry Jones had done even better than Thomas Taylor. With $46,000 in land value in surrounding townships as well as in Catlin, as the census taker estimated, he was the second-richest person in the township. His farmhand was from England.

The Sanduskys were richest of all. Josiah Sandusky had $110,000 in land value, and his brother Harvey had $45,000 worth, partly inherited from their father, the lecture sponsor Isaac Sandusky, who had died in 1852. Even though the family sold off some of their acreage to their invited English neighbors, they still had plenty left. Harvey Sodowsky, who refused to Anglicize his name, introduced shorthorn cattle and was an accomplished breeder. A good businessman, he profited further when he sold the English immigrants livestock to start their own cattle operations. Milton Davis, who had come from Ohio, was the fourth-richest person with an estimated $25,000 worth of land.

Through growing grain and raising cattle, these relatively wealthy landowners had acquired their riches without slaves. Thomas Taylor must have been fully aware of the North's advantages, and when war began, he was probably relieved he had left the South.

While skirmishes between residents sympathetic to one side or the other of the slavery debate were reported in Vermilion County, most residents, no matter from where they came, were intent on farming and getting their products to market.

The mid-1850s were a good time to begin farming. Prices were going up. England sought to import Illinois grain to support its military as it prosecuted the Crimean War, and railroads were expanding to get the grain to seaports. A correspondent writing on June 24, 1855, reported in the *New York Times* about Illinois: "[N]ever before in the history

Ambivalent Citizens

of this State, has one half as much prairie been broken—in any Spring—as this, most of which has been planted with corn." Wheat was also an important crop. Although other prairie and plains states would eventually surpass Illinois in wheat production, in the 1850s Illinois led the nation in growing the grain. It was ahead of other wheat-producing states in settlement and the amount of land already broken.

Before the railroads were built, Vermilion County's grain made its way by wagon to the Wabash, then by barge down the Ohio and the Mississippi Rivers to New Orleans to be sent around the world. Cattle and wheat, however, went overland to Chicago, as did apples and potatoes, to feed that fast-growing city and to be sent through the Great Lakes, the Erie Canal, and the Hudson River to New York City.

Illinois is now the second-greatest corn-producing state in America, after Iowa. Despite the plenty described in the *New York Times*, though, in the 1850s corn was less valued. It was grown mainly as feed for cattle and not, as the Illinois Corn Growers' Association bills it now, for "food, feed, fiber, and fuel." Nevertheless, east-central Illinois in the 1850s produced more corn than any other region of the state and almost seven times as much corn as any other grain. This was probably because farmers in this swampy region still raised more cattle than did farmers elsewhere in the state and had more need for feed.

Goods and services were arriving in Illinois at a rapid pace. Banks were few until 1856. In that year a private bank opened, survived, and became the First National Bank of Danville when Congress passed the National Bank Act in 1864. W. R. Woodbury, who had graduated from Rush Medical College in Chicago, bought a fledgling drugstore operation in Danville in 1853 and improved it. Abraham Lincoln was said to have been a customer.

Getting to Grand Prairie

In December 1854, a post office was established in Butler's Point. Signaling a new industry about to emerge, Abraham Lincoln's friend and law partner, Ward H. Lamon, established the Danville Coal Mining Company in 1855.

In the same year, the Illinois legislature gave Danville the county seat and a new charter. This established the city limits containing the original town plus such additions "as had been platted, or such as should farther be regularly platted and recorded as additions to it." That might have given Butler's Point residents the idea to plat their town, since it was now officially separated from Danville.

In 1855 the legislature passed the present free school system, giving all children, rich or poor, opportunity for education. Churches continued to proliferate, but the English held back from joining their American neighbors in worship. Emma Barker Dickinson, for example, was described in the 1889 *Portrait and Biographical Album, Vermilion County, Illinois* as a member of the Methodist Episcopal Church in England. It went on to say, "although she did not identify herself as a member here, she was an attendant on divine services and she lived an exemplary and Christian life." One wonders what made her reluctant to join, if she had been a Methodist back in Lincolnshire.

The biographical album was written in the late 1880s, when there was some perspective on the 1850s. But Emma's hesitation is another piece of evidence that the English immigrants liked to stick with their own kind, and religious practices were a way to do that.

Even into the late nineteenth century, the English still did not fancy their American-born neighbors as marriage partners, and the feeling might have been mutual. The descendants of the English-born immigrants of the

156

Ambivalent Citizens

1840s and 1850s continued to marry mostly descendants of other English-born immigrants.

Russell Puzey, the great-grandson of the immigrant Richard, remarked upon this curiosity in his family history. "We never intermarried with the Taylors," he wrote, naming the family that arrived in the Grand Prairie about the same time as his ancestors. "We were always good friends but never married." The Taylor genealogy, assembled by the late Josephine Taylor, shows it took until March 4, 1945, when Henry Jones's descendant Gwen Clark married Norman "Short" Taylor for a marriage between a descendant of one of the English families and a Taylor to take place. By that time the standoffishness of the 1850s had been forgotten.

—ɯ—

It wasn't just marriage that kept the English-born and their American neighbors apart. Several writers have commented on the tension between American-born settlers and English immigrants in Illinois at the time. Tensions, of course, still exist between American-born citizens and immigrants.

On the one hand the English brought skills, cash, and the affirmation that life in Illinois was worth the difficulty of uprooting themselves. On the other hand, historian Charles Boewe reported that "the Americans...thought the English were aloof and snobbish. The English...considered the Americans uncouth and boorish."

Tensions arose not only between the native-born and immigrants but also, in many places, among the immigrants themselves. The Grand Prairie English, however, had advantages that some larger and more famous settlements of English immigrants did not possess.

Getting to Grand Prairie

With a few exceptions, the Grand Prairie English were able to maintain good relations with one another. This was partly because there was no obvious dispute over a woman as there had been 135 miles south and forty years earlier. In Albion, Illinois, on Boultinghouse Prairie—also known as English Prairie—the community split because of the stolen affections of one Eliza Andrews. The families of founders Morris Birkbeck and George Flower, who ultimately married Eliza, didn't speak to one another for many years, and then it was only after Birkbeck had drowned.

Perhaps the Grand Prairie English owed their relative harmony to the absence of an overbearing, deeply felt ideology, a controversial religion, or an unrealistic goal. Such strictures tore apart the 691 Britons who between 1843 and 1850 established themselves in Mazomanie, Wisconsin, under the auspices of the British Emigration Temperance Society. While many of these immigrants were ultimately successful, the organization was soon disbanded after being riven by lawsuits and accusations. A few groups of socialists who settled in Wisconsin didn't fare well either.

16

Lords of the Soil

In the late 1850s, despite any doubts they might have had about their move to the Grand Prairie, the English-born immigrants plowed ahead—literally—toward mastery of their land and its products. The late nineteenth-century county biographies gave them credit for their scientific farming methods. They were experts in drainage. They planted orchards filled with nut trees, fruit trees, and gooseberries. They knew about "nitrogen fixing legumes" and the early clovers that were called "English grass." Their American-born neighbors surely knew about collecting manure for fertilizer, but none did it as reliably as the English. "The English farmers who hauled manure onto their fields raised the best crops in the region," said historian William E. Van Vugt.

The most difficult adjustment for all the English immigrants to Illinois was the "acceptance of a more primitive level of services—transport, roads, mills, forges, public houses, churches, schools, banks, and post offices—compared to the high standard in England," according to historian Charlotte Erickson.

Facing these conditions, the Grand Prairie English coped by sticking together, maintaining some of their old ways, supplementing their provisions by ordering tea and other treats from the mercantile relatives of the Puzeys back in England, and starting their own shops.

According to H. W. Beckwith's biographies in the *History of Vermilion County Together with Historic Notes on the Northwest,* in 1855 Henry and Sophia Puzey Church moved into the village of Butler's Point from their farm amid the other English three miles south. Henry established a store at a location eventually called Champion's Corner in the late nineteenth century. It was named after their daughter Jane's husband, Francis Champion, who ran the establishment after Jane's parents died.

This place name has mostly been lost, but it indicates a plot of land near the railroad tracks at the intersection of Paris and Vermilion Streets and the Lyons Road in present-day Catlin. The name is now mostly known by those who engage in an activity called benchmark hunting, in which participants track down surveyors' benchmarks embedded in the earth or marked in other ways. At Champion's Corner, the benchmark number is LB1841.

New businesses such as the one Henry and Sophia established increased the goods and services the settlers would have enjoyed. By the census of 1860 John and Sarah Taylor had abandoned farming and moved to Danville. John then continued his old craft as a blacksmith. Carrie Clipson Moore started a millinery shop in Georgetown.

Those who continued to farm found additional commercial pursuits. Henry Jones, of course, was one of the most active entrepreneurs. At some point in the 1850s, he, William Hind, and Thomas Bentley, who had been a tanner in London, established a tanning business. The tannery

Lords of the Soil

thrived, according to the Jones family history, until the people of the area became so busy with farming that they couldn't peel enough of the oak bark that was used in the tanning process.

In 1856 the family historian says Henry went into "the mercantile business with his son Richard, without, however, abandoning his farming interests." It is clear that on January 13, 1857, Henry paid $1,125 for groceries and fixtures to Frederick Tarrant, the second husband of his daughter Eliza.

That must have meant that Frederick and Eliza were running the store before Henry and Richard took it over. With shops run by the Tarrants, Joneses, and Henry Church, there would have been more opportunities for the neighbors to buy sugar, coffee, tea, and spices, the exotic groceries that had to be shipped long distances.

Acquiring merchandise for a store became easier in 1856. That is when the Great Western Railroad, also known as the Toledo, Wabash, and Western Railroad, was laying track south of Butler's Point. The company's president, J. M. Catlin, agreed to install a local station on the rail line. Community leaders were so grateful they renamed the village after him. Two years later, Catlin Township was formed.

With a new train station and a new name, the people of Catlin needed a good celebration, especially when the Fourth of July came around. So Henry Jones provided one. The Jones family historian, Genevieve Carter, read about it in Beckwith's history and reported it in this way:

It was one of Henry Jones's favorite desires to show these Yankees how they would celebrate such an occasion in England, if they had ever been so fortunate as to have such an affair there. He had been

161

Getting to Grand Prairie

brought up under the "lion and the unicorn" and had never been accustomed to see a Fourth of July and held to the traditions of his fathers that "St. George was a bigger man than ever the Fourth of July was." But on coming to America, he changed his mind, and became a thorough Yankee. To have the biggest celebration ever seen in the Wabash Valley was what the people of Catlin proposed...He was unanimously chosen president of the day...Twenty stalwart men were sent out who spent a week soliciting provisions. Wagon trains were pressed into service to bring in the abundance of the land. No such sight was ever seen until the commissary trains of the Grand Army of the Union took up the line of march in the sacred soil of Virginia...The fund of provisions was ample and the baskets full of fragments which they took up, but were never counted, but there was enough to keep Jones's hogs for weeks, after having given away to all the poor they could find.

Beckwith described it too: "Mr. Jones told them to go into his herd and slaughter all the fat steers they wanted. 'If a dozen won't do 'em, take a hundred,' said Mr. Jones. 'Give 'em enough to eat or they can't be 'appy'...Crowds of people came in from the surrounding country and father Jones was 'appy."

It sounds like a celebration the American-born population must have enjoyed. But they also might have viewed it as ostentatious or an overbearing Englishman showing off. The teetotalers might have attended, but also might have been scandalized by the English immigrants' tendency to view drinking alcohol as a normal part of any celebration.

Even now the people of central Illinois have a word for people whose behavior they don't quite approve of. They

Lords of the Soil

say they are "different." Different can apply to an idiosyncrasy, a baffling behavior, or an annoying habit. It is mildly pejorative, but it reserves harsh judgment. Whatever the American-born residents may have thought of Henry Jones's spectacle, it was the last celebration for a while.

—⁓—

By 1856 Henry Jones and his family had been in America for only seven years and other Grand Prairie English no more than nine. Henry's cemetery began to come in handy. The Clipsons lost a child, Matilda, who was born after Albert. She died of whooping cough at age four. The elder Thomas Bentley died a few weeks after the Independence Day celebration, leaving Eliza a young widow. Thomas Onley died in August. Then Henry Jones's wife, Sarah, died in September. James Bentley, Thomas's brother, died in February 1858.

Late in 1859 Henry Church died of the "tremens" after having been ill for six days. Several American-born children of the English immigrants died, including Frederick A. Church at three months old and Walter A. Church at seven months. Both succumbed to the whooping cough in late 1859 or 1860.

At some point after their spouses died, Eliza Hough Bentley moved in with Henry Jones. He was her uncle by marriage, and she and her children were living with him when the 1860 census took place.

—⁓—

With a new railroad station and a new official name for the village, the name for the undefined Grand Prairie would become moot. It took some time, though, before the name

163

Getting to Grand Prairie

"Catlin" rolled off the tongue. In an 1862 codicil to his will, Henry Jones still used "Butler's Point" as the name of his residence. Community leaders felt the village's improvements gave Catlin the chance to become an important resource for the state and perhaps the entire Northwest Territory. As early as 1850, all the land north of the railroad had been brought into cultivation. By 1858 all the land southwest of the station was taken and made into farms.

Catlin was less populated than Danville, but with the federal land almost gone, Danville's importance in the region declined. Population increases had caused outlying portions of the original Vermilion County to join or form other counties. By 1859, the county settled on the boundaries in existence today.

Meanwhile, Catlin civic leaders were planning for the future, filing plats showing proposed new streets and blocks and filling in the space between the almost forty-year-old Butler's Point and the new train station.

In 1857 Henry Jones's son Richard, now thirty-three, became the first station agent for the railway. He held this post for twelve years. He and Josiah Sandusky soon built a large warehouse, the precursor to the grain elevators of the twentieth century. They proceeded to buy and sell grain. All the while Richard continued to run a dry goods store, nominally with Henry, but within a few months, they had taken in the youngest Jones, Frederick, aged fourteen, as a clerk. By 1858 a surviving tax record for Catlin Township shows that Henry had acquired 1,268 ½ acres within that township alone.

The English immigrants were now buying lots in town. Henry and Sophia Church sold the Clipsons a lot for one hundred dollars. There is no evidence either couple ever lived there or built a house, but it certainly was a show of confidence in the village's growth and prosperity.

164

Lords of the Soil

During this time Abraham Lincoln would have been a familiar sight to the English. As a country lawyer riding between courthouses in central Illinois, Lincoln passed through the newly named village along the road on which Henry Jones's family lived. He was said to have been a regular patron of the Buckhorn Tavern in Salinas, a settlement west of Catlin now known as Fairmount. The name was changed because "Salinas" was already in use elsewhere in Illinois at the time Fairmount's post office were established.

In 1857 Lincoln had the duty of representing Thomas Sandusky, Harvey and Josiah's brother and the late Isaac's son, before a jury in the Vermilion County Circuit Court. Thomas had been indicted for selling whiskey without a license to Mr. Baum, Mr. Williams, Mr. Guyman, and brother Josiah. Sandusky pleaded guilty and was fined ten dollars.

The English immigrants might have followed the news about the organization of the Republican Party in the mid-1850s, and they might have read about or attended one or more of the Lincoln-Douglas debates in 1858.

Or they might not have done so. They probably weren't much interested in Lincoln's race for a U.S. Senate seat, which he lost, in 1858. They were only just beginning to consider citizenship. In 1859 Adolphus Church was one of the first to take that step, according to his report on the 1920 U.S. census. Henry Jones took out citizenship papers in 1860, and George W. F. Church did so in 1861. Few of the other English immigrants became citizens prior to the Civil War. None of the English-born women became citizens. After all, taking such a step was useless. They could not vote.

Getting to Grand Prairie

The late 1850s saw the arrival of the last of the English immigrants connected with the Grand Prairie English families—at least until long after the Civil War. John Carby and John and Richard Todd, with whom William Dickinson had lived in Lincolnshire as a boy, showed up. John Todd married Alice Bentley shortly thereafter. John Todd's occupation in the 1860 census was listed as grocery keeper, and he also had had enough cash to buy $225 worth of land. Thomas Puzey, Albert and Henry's brother and Richard Puzey and Sophia Church's nephew, arrived on May 15, 1857, according to his naturalization papers. The fourth Puzey brother, Jonathan, arrived about the same time.

Henry Lloyd, seventeen, came from Berkshire in 1858, according to his Vermilion County biography, and married Sarah Church in 1860.

—⚅—

The Grand Prairie English immigrants' success in their new land showed that, whatever unrealistic dreams they might have had before they arrived, they adapted to their new situations. They also felt they had more reason to succeed than their American-born neighbors.

"On these estates we hope to live much as we have been accustomed to live in England," said one English-born letter writer, whom historian Charles Boewe quoted. "But this is not the country for fine gentlemen, or fine ladies of any class or description, especially for those who love state or require an abundance of attendants.

"To be easy and comfortable here," wrote the immigrant, "a man should know how to wait upon himself, and practice it, much more indeed than is common among the Americans themselves on whom the accursed practice of

166

Lords of the Soil

slave keeping has, I think, entailed habits of indolence even where it has been abolished."

Another writer, also quoted by Boewe, claimed that English immigrants' habits destined them for success. Driven by the substance, high standards, and the well-kept appearance of certain English farms, the Englishman "plants himself squarely before his difficulties, he evades nothing but works hard and steadily to remove them." The same writer also had praise for Americans from the East but considered Southerners to lack industriousness.

The 1860 U.S. census revealed the success of various groups, at least when it involved land ownership. This census, unlike that of 1850, counted Catlin Township residents separately from Danville's. Catlin Township's population was 1,793. The total number of households in Catlin Township was 319. Of that number, 184 or about 58 percent owned land.

Some groups had done better than others. English-born men and women (including the Scot James Thompson, who had come with the Jones family) in Catlin Township headed nineteen households. Sixteen, or 84 percent, owned land. Most were clustered south of the village.

Ohioans had done well too. Of the seventy-one heads of households who had been born in that state, 60 percent were landowners. Two-thirds of the forty-three Kentucky-born heads of households owned land. Virginians, who were the fourth most populous group of heads of households, had done almost as well as the English-born. Seventy-five percent owned land. A few single men with small holdings lived as farmhands with other families.

Illinois was still a state filled with immigrants. Illinois-born residents were numerous because of the children born in the last two decades, but Illinois-born heads of households

167

Getting to Grand Prairie

in Catlin Township numbered twenty-two, slightly more than the English-born. Only three owned land.

The English-born group not only had a high percentage of landowners, but on average they also had more land. Of the nine English-born heads of households possessing at least $1,000 in land value, the average was $8,422. Only Kentucky-born landowners surpassed them in value, and their numbers were smaller. Eight Kentuckians had land with a value of more than $1,000. The average land value for those eight was $10,941.

Two Massachusetts-born landowners shared a total value of $23,400. Their numbers are too small, however, to account for a trend.

In Carroll Township, immediately south of Catlin, the 1860 census showed that the members of the younger English-born generation were doing all right for themselves in land too. Henry Puzey held $2,000 worth of land, Adolphus Church had $2,400 and Albert Puzey had $1,200 worth.

The English-born immigrants must have had mixed feelings about their situation in 1860. They had achieved their goals of land ownership and status, and they had maintained English ways in the midst of a vast, unfamiliar continent. In a 1959 letter to a distant cousin, Russell Jones, a descendent of both the Joneses and the Bentleys, reflected on the situation in Grand Prairie. "[They formed] an English settlement," he said. "All they needed was a queen."

Though the Grand Prairie English might not have missed the soupy skies of London, some must have longed for the city's advantages. Diarist Eliza Jones Browne Tarrant especially must have been homesick. How did she manage without the theater and her excursions? Henry Jones and Will Clipson surely missed the pubs, horse racing,

Lords of the Soil

and betting. They might have missed the history, cultural stability, and perhaps even the architecture. Church buildings, for example, in rural Berkshire and Lincolnshire, two English counties with which many immigrants were familiar, are small but well-appointed and solidly built of stone. A prairie landscape without such evidence of thousands of years of human activity might have looked bleak in comparison.

—w—

By the end of the decade, some members of the Grand Prairie English were dispersing, seeking additional economic opportunities, although many returned. Frederick Jones went off to Lafayette, Indiana, and then to Indianapolis as an apprentice to a blacksmith before he returned to help his father in the store.

In 1860 Eliza Jones and her second husband, Frederick Tarrant, were living with Eliza's daughter, Emily Browne, and their daughters, Miriam and Sarah, in Detroit where Fred was working as a laborer. Fred's shipmate, Isaac Winch, and other Berkshire-born immigrants who had come with Fred on the *Cosmo* were living nearby. The Tarrants had left Illinois after Fred sold the store equipment and merchandise to Eliza's father. At last Eliza was living in a larger community where there would be more entertainments. By 1870, though, the Tarrants had moved back to Catlin Township.

John Swannell and his second wife, Jane Clipson, moved to Leavenworth County, Kansas, where she died in 1859. He then moved back to Vermilion County with his daughter, Eva, by his first wife, and son, William, who was Jane's.

The dream that land was the key to status, independence, wealth, and happiness had become real for many of

169

Getting to Grand Prairie

the English immigrants. Land in America, though, came with a price that would reveal itself vividly in the 1860s.

The English novelist Anthony Trollope said in his novel, *The Way We Live Now*, "Land is a luxury, and of all luxuries is the most costly."

17

The Americans' War

In November, Abraham Lincoln was elected president of the United States. He took his leave from Vermilion County the next February when he stepped from his train onto a platform in Danville and said if he had any "blessings to dispense, he would certainly dispense the largest and roundest to his good old friends of Vermilion County."

Lincoln took office in March, by which time seven Southern states had seceded from the Union to form the Confederacy. The Civil War began in April. Soon President Lincoln called for three hundred thousand men.

Like the English-born in other parts of America, the Grand Prairie English men reported quickly for duty. They did not form ethnic regiments as did German and Irish immigrants, but signed up along with their American-born neighbors. Ultimately more than fifty-four thousand English-born men served in the Union army, a high percentage compared to other immigrant groups.

The high numbers partly reflected the Englishman's distaste for slavery. However, the English-born immigrants also might have felt a need to prove their loyalty, since there was still a certain amount of residual anti-British feeling

Getting to Grand Prairie

from the War of 1812 and from a fresher western boundary dispute between Canada and the United States that was
resolved in 1846. As reassurance to their American neighbors, some English-born farmers painted American flags
on their barns to show their solidarity with the Union. As
landowners the Grand Prairie English would have felt their
stake in America worth fighting for.

The first of the Grand Prairie English to go off to war
was cabinetmaker and twice-widower John Swannell, aged
thirty-two. On April 18, 1861, he and several neighbors joined
Danville's Company C of the Twelfth Illinois Infantry, the
first Vermilion County company to be formed. This was six
days after Confederate forces attacked Union forces at Fort
Sumter. It is not known how he arranged for his seven-year-
old daughter and infant son to be cared for.

Twenty-year-old Billie Clipson, a painter and decorator, was the next to go. The family story says he and his
father were disagreeing one morning at the breakfast
table, and Billie angrily left, picked up milk pails and
walked to the barn as if to milk. He didn't return. Instead
he joined Company A of the Twenty-first Illinois Infantry,
which was organizing in Decatur, on June 15. Colonel
Ulysses S. Grant led the company until August 7, and the
family story says Billie was one of Grant's favorites. Philip
Pusey, who still sometimes spelled his name with an "s,"
and Richard Todd joined Company I, Thirty-fifth Illinois
Infantry, on July 3, 1861, along with twenty-four other men
from Catlin Township.

Benjamin Onley enlisted in Company A of the
Seventieth Illinois Infantry Regiment in July 1862, but
was mustered out in October of that year at Camp Butler
in Springfield. Jack Clipson, a carpenter, joined Company
D of the 125th Illinois Infantry on August 10 and left in

172

The Americans' War

September. He wasn't alone either since his unit was made up of men "drawn from the rural precincts of Vermilion and Champaign counties," according to the Illinois adjutant general's report, now online in the same archive that details each company's roster and history. His company was filled with men mostly from Georgetown, a township in which Jack probably had friends since the Clipson farm was on the Georgetown-Catlin border.

At the same time, John Todd and Henry Lloyd, the only married English-born volunteers, joined Company G of the same unit. John's occupation was listed as a grocer on his Civil War papers, and Henry's was a butcher. The thirty-one-year-old Thomas Bentley went with them, as did Thomas Puzey, who later transferred to the Sixtieth Illinois Infantry.

By the spring of 1863, the Union had a draft. Thirty-one-year-old Jonathan Puzey joined Company F of the Forty-fourth Infantry in September 1864. He served only about nine months before his regiment was disbanded. Benjamin Onley, now listed as living in Georgetown, enlisted again in March 1865, this time as a musician, in Company G, Twenty-eighth Illinois Infantry Regiment, and served for a year.

The Grand Prairie English-born infantrymen were only eleven out of dozens of volunteers from the township, the population of which was almost two thousand in 1860. Interestingly no Church or Jones men volunteered. One suspects that the attitude in these unceasingly entrepreneurial families was that this was an unnecessary war that interfered with their goals of prosperity, status, and independence.

Another reason was that most of the Church and Jones men had families. The young volunteer farmers tended to be unmarried with fathers or grown brothers at home who could continue farming. After all, wars don't obviate the

173

Getting to Grand Prairie

need for food. Among other resources, the North's ability to produce food for its army and its populace was a reason it won the war.

While the eleven English-born sons and their fellow townsmen were trudging through mud in Tennessee and Georgia, those back at home were beginning to venture into leadership roles in business and civic matters. George W. F. Church was elected clerk of the township but not before he found time to return to England for a visit with his mother and father. He returned to New York on April 3, 1861. During the war Richard Jones was named township supervisor with William Hind as clerk. Richard continued to serve as supervisor intermittently until his death.

William Hind, who had been a printer in London, sank the first coal shaft in 1862 in Catlin Township, about a half mile west of town. It operated for only eight years, but it was a precursor of the coal mining operations that lasted for about sixty years, some of it on Clipson and Puzey land, beginning in the late 1800s.

The event that would have rocked the town in 1862, however, was not the sons going off to war, and it wasn't a coal mine. It was Will Clipson's suicide.

Suicide was unusual in the Grand Prairie, but if it were going to happen, an English immigrant would have been a more likely candidate than some others. A study in the early twentieth century in New York and Boston showed that English immigrants committed suicide at about the same rate as French immigrants, but twice as much as the Irish and three times as much as Italian immigrants. Germans, who had not yet settled in east-central Illinois in significant numbers, were twice as likely to commit suicide as were the English. Protestants like Will Clipson were more likely to kill themselves than

174

The Americans' War

Catholics, probably because the strictures against suicide were and are stronger in Catholicism than in mainstream Protestant religions. Generally in the mid-nineteenth century, there was a higher incidence of suicide—almost three times greater—among American immigrants compared to the native-born population in the countries they had left. Loneliness, loss of family and a familiar culture, and difficulty adjusting to a new life all must have factored into this tendency.

Later Clipson family members called Will a "misfit." He had not chosen to be a farmer in England, and they suspected he didn't enjoy farming in Illinois. A review of suicides at the time in England showed that men over fifty-five were more likely to commit suicide than younger men. The authors attributed this situation to the greater likelihood of illness and the humiliations of old age.

It is possible Will Clipson suffered from an illness of which his family members were not aware. There is evidence for depression and loneliness, given his reluctant but speedy leave-taking from London and a possible longing for his old life. Fear for his sons' safety might have played a role too. The timing, a month before a second son was scheduled to go to war and three weeks before another would be eligible to sign up, gives support to the family story that he killed himself, at least in part, to save a son.

Another shock to the town was Henry Jones's death on November 10, 1862. A driven, resourceful man, Henry, fifty-eight, weighed three hundred pounds and was possibly weighed down by a heavy heart. Four children, his wife, and Will Clipson were gone. But he had had solace in his last years. His late wife's niece, Eliza Hough Bentley, had moved in with him after Sarah Jones and Eliza's husband, Thomas, had died.

175

Getting to Grand Prairie

Eliza Bentley must have been a comfort to him, but perhaps she was more of a comfort than the family story recognized or admitted. On September 5, 1857, in his first will, he appointed his son Richard to administer a trust fund of two thousand dollars on Eliza's behalf for her lifetime. Fellow immigrants James Thompson and George Vandersteen witnessed this will.

On his deathbed, however, he reconsidered.

It's easy to imagine a dramatic, Victorian-era scene with Henry gravely ill, his friends William Morris and William Millimore, two later-arrived Englishmen, and his son Richard beside him, and Eliza sobbing in the next room. That day he must have felt he hadn't done right by Eliza. He had his friends prepare a codicil to his will that Morris and Millimore signed. He revoked the trust handled by Richard and instead left Eliza a life estate in the east half of the northeast quarter of Section 34, Township 19, Range 12. He had bought this land at two different times. Neither transaction records the acreage exactly. Instead both times the deed refers to the land only as "part" of the east half of the northeast quarter of the section. In 1851 Henry had paid $90, and in 1856 he had paid $250 for these parcels.

Upon Eliza's death, he declared, the land was to pass to her children Alice, who had married John Todd, and James Bentley. He did not mention Eliza's other surviving children, Henry and Jessie. Then came what might have been astonishing news or a secret badly kept. Henry's will says, "the above bequest unto said Eliza Bentley and unto her said children is not to debar them or any of them from sharing my other property with my other children equally." My *other* children?

Henry's codicil continues: "The said Eliza Bentley or her said children, in addition to the above bequests in their

176

favor, are to share in any other property real and personal in the same manner as intestate estates are distributed under the statutes of Illinois now in force."

There are not many ways to interpret this. Henry was declaring that Alice Bentley Todd, born in London, and James Bentley, born in Illinois, were *his* children, not the children of Thomas Bentley, Eliza Hough Bentley's husband. It was obvious from Eliza Jones's diary and Henry's shipboard letter that the Joneses and the Bentleys were close in feeling and location, and Eliza and her children moved into Henry's house shortly after their spouses died. Perhaps their relationship was known to many, or maybe it was a secret.

Notably Henry did not leave land to Eliza's other two surviving children, whose father, presumably, was actually Thomas Bentley. In addition to receiving the land Henry left to Eliza in a life estate, Alice and James Bentley were to share equally with his "other children" in the rest of his estate.

The change in the will improved Eliza's situation considerably. It gave her control over the land during her lifetime. She wouldn't have to rely on Richard, whose opinion about this matter remains unknown, to parcel out her income from the trust. It also probably gave her a better income. This would have been especially true during the Civil War, when prices for grain and cattle were high. The land itself, if valued at the thirty-five dollars an acre Henry paid Will and Matilda Clipson in 1857, would have been worth at least $2,800, which was, of course, already more than the $2,000 cash investment Henry had originally settled on her. Finally it gave two of her children an inheritance from Henry, both in the land they would receive at her death and in the division of Henry's wealth shared with his "other children."

The family story ignores this interesting matter, or perhaps the family historian didn't look at the will or notice

Getting to Grand Prairie

this part of it. Henry's admission might or might not have affected Eliza's reputation, but the deed was done, and he was gone. His son Richard moved into the family home.

—⁂—

In 1863 the leaders of Catlin incorporated their village, showing some confidence in the success of the Union in the battles still raging in the South.

The English families kept in touch with their London relatives, who were worried about them. In January 1864, Fanny Bentley Church's aunt, Fanny Holland Palmer, the sister of her late mother, Sarah Holland, wrote her a long letter. Aunt Fanny was worried that Fanny Church's husband, Dolf, or A.B., as he was known in Illinois, might have to join Fanny's brother, the younger Thomas Bentley, at war. Adolphus, however, stayed home.

The letter is full of domestic doings. Aunt Fanny's son William had been married three months before. Aunt Fanny's daughter, also Fanny, was growing up and sent a picture of herself. Aunt Fanny mentioned the London Church family, whose pre-emigration connection with the Bentleys and Joneses was only conjecture until this letter revealed it. Aunt Fanny described receiving a letter by Mr. Church, Fanny's father-in-law, George Zephaniah. Aunt Fanny declared she had never met Fanny's husband Adolphus, but she had met his brothers George and Albert. She might have met George when he returned for his English visit, but she had to have met Albert before he left England in 1850. Therefore, the Church family and the Bentley-Jones families knew one another in the 1840s in London before the families emigrated.

178

The Americans' War

Remarkably Fanny Bentley Church's seventy-nine-year-old grandfather, William Holland, was still alive, although Aunt Fanny wrote he was feeble. Aunt Fanny said she would write to Fanny's older sister Sarah, the wife of Richard Jones.

—⚉—

Aunt Fanny Palmer's fears about the safety of Fanny Bentley Church's brother were well founded. A little more than a year later, on March 4, 1865, Thomas Bentley's funeral was held at the home of Edward Wingrove on the "Dickerson Pike," a road name now lost. Thomas had been discharged in Nashville a year before and died there, according to the later obituary of his stepmother, Eliza Bentley.

Other English-born soldiers did not fare well either. When he became the first to enlist, John Swannell was a widower twice over with two children, but he was living in a Danville boarding house with other single men and women. John Swannell might have had little to gain by joining an American war, but he also might have felt he had little left to lose. He lasted less than a year in the war. Wounded at Fort Donelson in Kentucky, he died in southern Illinois in June 1862.

Philip Pusey had spent the early months of 1863 sick in bed in Missouri, according to his company's muster rolls. When he revived he put his blacksmithing skills to work for his regiment. But he spent two more months sick with a respiratory infection that affected him the rest of his life.

Billie and Jack Clipson survived but barely. After three years, with Billie's unit participating in battles in Perryville, Murfreesboro, Chickamauga, and Atlanta, the whole unit

179

Getting to Grand Prairie

was reenlisted at Ooltewah, Tennessee. Billie, though, was not physically present.

He had been taken prisoner. The family story quotes his great-nephew Roy, saying that Billie was in Andersonville Prison for eighteen months. Prison records prove it. Chickamauga, occurring in September 1863, was the most likely spot of his capture. His unit suffered many losses, and many men were captured in that battle, which the Union lost.

Appalling conditions at Andersonville prompted Billie to attempt an escape on December 7, 1864. Walter Hartsough, of K Company, Sixteenth Illinois Infantry, told the story to surgeon John J. McElroy, a member of the 125th Regiment of the Illinois Volunteer Infantry. McElroy would have been especially interested because he was also from Catlin Township and had spent time in Andersonville Prison. In 1879 McElroy published a book, *Andersonville: A Story of Rebel Military Prisons*, in which he recounts Hartsough's tale.

Hartsough, Billie Clipson, and a man named Frank Hommat of Company M were part of a group that had been taken to Thomasville, Georgia, and were about to be returned to Andersonville. The men decided to sneak away.

They stored food for their escape, ate heartily to gear up for the venture, and stole past guards camping in the woods. Each heading a different way, they proposed to rendezvous in a small swamp with a stream from which they had previously taken drinking water. Hommat and Hartsough waited for Billie. They submerged themselves in the swamp under a log, but he didn't come. So the two men went on, eating turnips from a field. Elderly slaves fed them and put them up for the night as did a widow from Massachusetts whose late husband had owned slaves. According to Hartsough's account of the events, "Our friend Clipson, that made his

180

The Americans' War

escape when we did, got very nearly through to our lines, but was taken sick, and had to give himself up. He was taken back to Andersonville and kept until the next Spring, when he came through all right."

Billie Clipson was released at the end of the war on April 5, 1865. The family story says he was in a sorry condition when he arrived in July at his widowed mother's house. But he still had a sense of humor. He walked into the house with the milk pails he had taken when he first left. "Mother," he is supposed to have said. "Here is your milk." Matilda apparently fainted.

Jack Clipson also lived through the Civil War as a scout with General Sherman's forces, although he was wounded near the close of the war, according to his 1924 obituary.

Their younger brother, James, who had been about to turn eighteen just before Jack left, never went to war. His father's suicide caused him to be the only man left on the farm.

Philip Pusey was with his company from the day it was formed until it was disbanded. Shortly after he returned from service in 1864, he married Amanda, daughter of Joseph Anderson, another English-born Union soldier who had come from Oxfordshire to Vermilion County about a decade before the Grand Prairie English arrived.

Richard Todd didn't make it. He was wounded at Chickamauga and returned to Catlin, where he died in April 1864.

John Todd also was killed in 1864. He left his wife, Alice Bentley, with two sons. Four years later, Alice married John's friend, John Carby, who had grown up on Frith Bank only a few steps away from the homes of the Dickinsons and the Barkers in Lincolnshire. In the 1860 census, John Carby had been listed as a farmhand on Alice and John Todd's farm.

181

Getting to Grand Prairie

Henry Lloyd made it back from the war. He was with Sherman on the March to the Sea and was discharged on June 9, 1865.

The village commemorated the war and its losses by naming one of its streets Lookout Street, after Lookout Mountain, near the battlefield of Chickamauga that had figured heavily in the movements of several infantries filled with Catlin men.

The soldiers who survived returned to a county that had seen its most important function die. Illinois was no longer the magnet for English immigrants it had been fifteen years earlier. By 1870 such regions as Utah, in which English immigrants amounted to 18 ½ percent of the population, held more promise. This was partly due to Mormon missionary work and also because cheap land was now farther west.

The land office, which had been opened in Danville in 1831 to oversee all federal land sales in its district, had closed in 1856. There was no need for it anymore since the federal land in the region was now in private hands. Records were sent to the Springfield office, which tied up any loose ends. New railroads had been built, and this reduced the importance of Vermilion County's navigable waterways for getting grain and cattle to market.

Champaign County, with shallower rivers and less timberland than Vermilion County, had seen little growth until the 1850s, but it was now desirable. The first reason was the Illinois Central Railroad, which connected Champaign, or West Urbana, as it was initially called, with Chicago and points south. The second was Lincoln's 1862 signing of the Morrill Act, which gave states thirty thousand acres of federal land for each of the state's members of congress and senators. States could sell the land to finance public universities.

182

The Americans' War

Urbana in Champaign County was chosen as the site of the incipient University of Illinois, which opened in 1868. Eventually Danville got a soldiers' home, which turned into a Veteran's Administration Hospital. It also had coal mines, railroads and a few growing industries that sustained it into the twentieth century. Vermilion County's heyday, in which land was the commodity, had disappeared. The dreams were no longer about land. A university education better fulfilled dreams of status, wealth, and a satisfying life.

18

A Legacy of Gooseberries

By the 1870s, it had become clear to the Grand Prairie English they were never going back to England. The immigrants began taking out citizenship papers in earnest and taking leadership positions in the community—something they had held off doing at first.

In October 1876, Thomas Puzey, Arthur Jones, Frederick Jones, and Henry Bentley became American citizens within a few days of each other. John Carby vouched for Thomas Puzey, so Carby must have taken the step earlier. Richard Clipson joined them as a citizen in November. James Clipson became a U.S. citizen in October of the next year. George Vandersteen took out his papers in April 1878.

In 1870 Albert Church became the village postmaster. Later James Clipson and his wife, Clarissa Douglas, became stockholders in a bank that merged with another bank that Richard Puzey Jr. started. James Clipson also served on the board of supervisors of Vermilion County.

By 1875 English-born men held three out of the four Catlin township offices. Richard Jones was the supervisor again, Frederick Tarrant served as clerk, and Henry Lloyd was the collector. In 1876 Albert Church replaced Frederick

Getting to Grand Prairie

Tarrant, and Richard and Henry stayed on. Richard Jones, aged fifty, died in October of that year at a birthday party for his sister, Sarah Jones Church. Nevertheless, leadership of the town stayed in English-born hands throughout most of the rest of the nineteenth century.

Led by a group of men who had crossed the Atlantic for the status and wealth that land promised, residents of the village of Catlin had dreams of glory. At one point they proposed that the county seat be moved to Catlin, since Danville had served its purpose with the now-closed land office and would no longer be critical to the century-old nation's growth. The prognosis about Danville's importance proved correct, but Catlin's growth was about the same as the larger town's.

In the 1900 census, Catlin Township's population was 2,207. It was forty-nine square miles with 102 rural homesites and 630 farm parcels. The geography of the township is about the same now as in 1900, as is the population, which was 2,087 in 2010.

The English-born and their descendants continued to marry other English-born Grand Prairie residents. Frederick Jones married Harriet Ann Dickinson, daughter of Emma and William Dickinson, in 1866. In 1870 Arthur Jones married Emma Dickinson, Harriet Ann's younger sister. Elizabeth Sarah Church, daughter of George W. F. Church and Sarah Jones, married Richard and Amelia Puzey's only son, Richard. Richard Sr. lived with them, and they farmed together, tending orchards filled with apples, peaches, cherries and gooseberries.

Some of the English immigrants' energy went to preserving English ways. Hattie Clipson and Eliza Tarrant were heralded for their hot cross buns, which they produced for many celebrations. Puzey family papers detail the orders

186

A Legacy of Gooseberries

the English-born residents of the Grand Prairie sent for tea and other English goods from their relatives in London who still ran a grocery.

But it became harder to behave like English men and women when English society was surely different from the way it was in the 1840s. Furthermore, the children of the immigrants were growing up with American, not English, influences.

A few of the English began attending the American churches, even though they were still reluctant to join. Emily Jones Church was one of the few who joined a Methodist church, which she did in 1873.

Two of the Grand Prairie English actually started a church. Richard Puzey, who lost his wife, Amelia, in 1867, and Matilda Clipson, Will's widow, were good friends. They had been next-door neighbors since 1853. The village of Catlin, only three miles away, already had a Methodist church, but Richard and Matilda decided they needed their own.

They and some neighbors, not all of whom were English-born, founded the Fairview Methodist Church in the late 1860s. In 1876 they built a rectangular church building with a gabled roof, a door, two windows in front, and three windows along the sides. It had no steeple. Matilda and Richard's neighbor, Benjamin Small, donated the land on which the church was built.

Matilda's son James was one of the first trustees. Some Puzey descendants believe that Richard was a lay preacher when he immigrated, and he filled the pulpit on some occasions, but "Mother" Clipson did too.

The original building burned, and in 1896 church members replaced it with the plain clapboard country church that

187

Getting to Grand Prairie

is still there. It does have a fair view, undramatic but as lovely and spacious as the view from Tom and Dee Belton's farm.

Mark Learnard is a church member who lives kitty-corner from the church and is a descendant of the Grand Prairie English. He said the church has only about twenty active members now. The original walnut pulpit has been preserved and refinished, made possible by a donation from Richard Puzey's great-grandson Russell. The church is plain on the inside—somewhat like New England Congregationalist churches of the early 1800s. Only a few of these Illinois country churches are left, saved from demolition by the tender care of people like Mark Learnard. The building's old-fashioned modesty is appealing.

Methodism worked out well for the English-born. Their Anglican faith had been supported by a clergy and traditions that were hard to maintain in America. Methodism emphasized traits helpful in a pioneer life—honesty, frugality, simplicity, and healthy personal habits.

The original immigrants who were still alive were active members of their community. One senses, however, less purpose as well as less isolation among members of the younger generations. This group began to marry American-born partners with increasing frequency. Henry Bentley, son of Eliza and Thomas, married Serena Goings in 1871. James Clipson married the Illinois-born Clarissa Douglas in 1876. The younger Joneses and Churches married men and women from American families more frequently as time passed.

The English-born families and their descendants continued to acquire land and build businesses, but the cash advantage they had had in the 1850s became less pronounced as the nineteenth century wore on. Only a few new English settlers arrived in Vermilion County after 1860, but

A Legacy of Gooseberries

American-born settlers continued to arrive in great numbers, and they prospered even more than the English had done a few years before.

Most of the English families had achieved their dreams as property owners and were relatively well-off. After the Civil War, however, they were not among the county's largest landowners. Vermilion County was known in 1870 for its large-scale farms on which farmers raised more stock than grain. Twenty-three farms in the county contained more than a thousand acres.

The 1870 census of Vermilion County indicates the value of the household heads' holdings. The Sandusky/Sodowsky families continued to stand out for their wealth. Several family members owned at least $50,000 worth of land. Men with holdings valued at more than $25,000 included the American-born James Sconce, Valentine McNeer, David Fisher, Charles Baum, Alvin Hildreth, Robert Bennett, Thomas Taylor, Jesse Davis, and Wilson Burroughs.

Richard Jones's land was valued at $16,000, and George W. F. Church's at $10,000. Other English-born landowners made do with significantly less wealth. Henry Puzey's land value was estimated at $3,930, an unusually specific amount. Other English-born landowners' holdings ranged from $800 for William Hind to $4,000 for Thomas and Louisa Jones Church to $6,000 for both Frederick and Harriet Dickinson Jones and Arthur and Emma Dickinson Jones.

The Grand Prairie English would have been familiar with the English system in which, among most landowning families, the oldest son inherited all. That kept land holdings large.

In America, though, all daughters and sons typically inherited real property equally. This meant the land was

Getting to Grand Prairie

perpetually divided. Perhaps there wasn't enough farmland to go around for the children of the original Grand
Prairie English, for some of them dispersed, mostly to
points west. Thomas Church and Louisa Jones married in
1861 and had moved to Kansas by 1880. The younger George
Vandersteen moved with his family to South Dakota in the
1880s. After the war Billie Clipson moved to Iowa, where
he continued his trade as a painter and decorator. He married the American-born Marintha Tipton and frequently
sent his children to visit their Grand Prairie grandmother.
In the 1890s, he took advantage of the Homestead Act and
acquired a farm in Boonville, Missouri.

Jack Clipson married Elizabeth Fairchild, a daughter of
early Illinois pioneers. For a time they lived with their son,
Edwin, in Fairmount, the farming community just west of
Catlin. The 1870 census shows Jack owning $2,000 worth of
land. But they divorced, and she married someone else.

He remarried as well. He moved to Iowa and wed
Margaret Hutchings. In 1880 his son Edwin was living with
them, and Jack worked as a carriage maker. According to
the 1910 U.S. census, he was living in Okmulgee, Oklahoma,
with his wife and their daughter, twenty-four, and working
as a salesperson in a dry goods store. He died in Okmulgee
in 1924.

The Puzeys, a family who had faced friction in England,
quarreled again. Albert and Henry had a falling out that
family members attributed to Albert's lack of success and
Henry's arrogance about his own prosperity. Letters poured
into Illinois from Berkshire pleading with Albert and Eliza,
who was accused of being meddlesome, to restrain themselves. In a fury, though, Albert sold his land, valued at
$1,200 in 1860, to his brother Jonathan and picked up stakes,
The family moved to Fulton County, about a hundred miles

190

A Legacy of Gooseberries

away. There the couple could be found in 1870 with their three children, Anna, Mary, and Fred. Albert was working as a coal miner.

At some point, Albert's wife, according to the Puzey papers, "kind of went bonkers, and they got a divorce, and she ended up in the Kankakee sanitarium." Eliza, aged fifty-eight, who had worked briefly giving music lessons, was committed on July 2, 1888, to the Illinois Eastern Hospital for the Insane, the Kankakee institution that had opened in 1879. Albert and Eliza's daughter, Anna Puzey Love, was appointed her conservator. Puzey descendants said that prior to Eliza's commitment, Albert had run off with another woman and Eliza had obtained a divorce in 1884 on grounds of desertion. Albert moved to Kansas. In 1890 Eliza's doctors declared her sane, and she went to live with Anna, who is also known as Sarah Ann.

According to the 1895 Kansas state census, Albert, aged sixty-seven, lived in Sawyer, Kansas, with his wife, Caroline, listed as C. A., aged sixty-three. Lizzie Blanch, nineteen, is living with them. Family letters show that Lizzie is Albert's cousin's daughter, whose parents and sister had been living on Albert's farm but had died within weeks of one another. Albert and Eliza continued to quarrel. This time it was over land their son, Fred, left when he died in 1887. They eventually split ownership of the land. In the 1900 census, the same Albert, now a widower, was living with Francis Blanch, twenty-five, who was listed as his niece but probably was Lizzie, the same cousin who was counted in 1895.

Thomas Puzey, brother of Albert, Jonathan, and Henry, had a sad life after the Civil War. He never married, and in August 11, 1873, he rescued a woman who was about to be struck by a train in Decatur. In doing so he was himself struck and lost an arm and a leg. He was estranged from

191

Getting to Grand Prairie

his family, and when he was admitted to the Soldiers' Home in Danville in 1902, he named Henry Lloyd, a friend, as his family contact. He died on May 19, 1905, struck and killed by a train on the Wabash Railroad. Family members suspected he had been drinking. One wonders if he, like Will Clipson, had attempted suicide and finally succeeded.

At some point after 1870, Jonathan Puzey left Vermilion County and moved back to England. In 1889 he visited his family in Illinois and Kansas for about a year, but then he returned to Stanford in the Vale, where he lived on his Civil War army pension.

The Thompsons moved on too after about ten years in Vermilion County. They went to the rapidly growing Champaign County and eventually to Chicago where James worked as "a foreman of John B. Sherman's fancy stock barns." Along the way he became a Mason and a Methodist. He died in 1890, and his body was brought back to Catlin to be buried in the Jones Cemetery.

Most members of the Onley family and their cousins, the Ansteads, gradually relocated to Georgetown or nearby Indiana. The younger Rachel Onley and her husband, John Cannon, moved to Covington, where he kept an inn in the 1870s. Rebecca Onley and her English-born husband, Edward Hough, left for St. Louis, where other Houghs had settled.

Held by the land, though, most of the Grand Prairie English remained in Vermilion County.

Eliza and Fred Tarrant ran Eagle House, a boardinghouse in Catlin Village. Fred also ran a grocery, and Eliza bought and sold small parcels of land, speculating all her life.

Three of Matilda Clipson's children, Hattie, Richard, and Tal, remained with her. She never moved to the lot in Catlin she and Will had bought in the 1850s. She stayed on

A Legacy of Gooseberries

the farm and seemed to lead an accident-prone life. Her obituary said she once broke a limb when a sheep attacked her as she was passing through a pasture. She also fell through a trapdoor and seriously injured herself. Her funeral service in 1901 was held at the Catlin Methodist Church rather than the Fairview Church she and Richard Puzey had started.

James Clipson and his wife, Clarissa, moved to a farm west of Catlin that had been in her mother's family, the Burroughs. Four big box elder trees, cherry trees, apple trees, and gooseberries surrounded the commodious house they built.

Richard and Tal Clipson were in business together. They shipped more than 11,000 hogs, 325 heads of cattle, and 1,050 sheep to Chicago in 1888, earning more than $108,000, according to the "Catlin Clack" newspaper column.

Tal married Ethlen Sanford of Hoopeston, a town in the northern part of Vermilion County. Richard lived with his mother, Matilda, who called herself Anna on the 1870 census form.

Richard was gay, said one of his brothers' descendants. Catlin's newspaper often mentioned his bachelor status. One example is this ditty written by "Hannah Mariah," the pen name of the "Catlin Clack" columnist G. Wilse Tilton, about a "phial" of whiskey someone provided for him at a party:

> Dear Friend Dick,
> If you are lonesome and sick
> And want relief quick
> Take a "drap" of this mint:
> But a permanent cure
> If you want to be sure
> A good wife secure.

In 1892 Hannah Mariah chided "Dick, a little bachelor of some twenty-seven summers" for ignoring a bunch of young women who descended upon his house.

While Hannah Mariah's column frequently teased Richard Clipson about his lack of interest in women, there was no meanness about it. It seems as if "don't ask, don't tell" worked well for these close friends and neighbors.

For a time Matilda and Dick lived with George Hines, a "fresh-air boy," from Chicago. Matilda said she adopted him, and Richard called him his ward. George is buried in the Clipson family plot in Oakridge Cemetery with Richard, Tal, Ethlen, Hattie, Matilda, and Will, whose body was moved from Georgetown. For some reason, Richard Puzey's son, Richard, became upset at the Jones Grove Cemetery, so Puzey descendants are also buried at the more diverse Oakridge.

George W. F. Church, as his father did in London, kept a large fishpond, which Hannah Mariah estimated held thousands of fish, some as old as six years. After his wife, Sarah Jones, died, he married Edwina Mary Church. She was twenty years younger, and her family was American with no relation to him. They had a child, Edith.

Little is known about William and Elizabeth Hind or John and Sarah Taylor. These childless couples had no descendants to keep their letters or family mementos. Their stories are lost.

—⁂—

Travel was getting easier, even in the middle of America. In 1877 George Vandersteen returned to England to fetch his older brother William, whose wife had died. In 1888 Richard Clipson took a tour through Iowa and Kansas, and "may visit the Pacific coast before his return," according to an otherwise undated newspaper account by Hannah Mariah. John Cork, a nephew of Amelia, Richard Puzey's late wife, came from the island of Jamaica to visit his uncle by marriage. Adolphus and Fanny Bentley Church frequently visited their son in Frankfort, Indiana. Arthur Jones and Albert Church went to Kansas when they heard of the serious illness of Louisa Church, Arthur Jones's older sister and Albert Church's sister-in-law. The trips are duly recorded in the "Catlin Clack" newspaper reports.

Even the Sanduskys, some of whom had moved farther west, returned to Catlin for a visit. In 1888 an Edward Hough, described as an English relative of the Joneses and Churches, visited his Catlin cousins.

Remarkably, a few stragglers with connections to the Grand Prairie English appeared in the latter part of the nineteenth century. William and Eliza Carby Dickinson immigrated with their nine children to Catlin from Lincolnshire in 1881. Eliza was the older sister of John Carby, who had married Alice Bentley Todd after her husband was killed in the Civil War. It is likely this William Dickinson was related to the Dickinson family who immigrated with the Clipsons, but the connection cannot be proven.

William and Eliza's son, John Arthur Dickinson, ran a grocery store in Catlin for many years around 1900. He dealt in "confectionaries" and also "conducts a lunch counter," according to Hannah Mariah.

The Churches kept in touch with their English and Brazilian relatives by frequent letters, which American descendants saved. The letters are full of news about the health and condition of various family members and the comings and goings of relatives in Australia, India, Scotland, and the East Indies. Especially touching are those from Adeline, the Church brothers' sister who had married the Portuguese-born Francisco Pereira and moved to Brazil. She described her growing family, her worry about her husband's business, their lack of money and servants, the family's poor health, and her own isolation, so far from her parents, brothers, and sisters.

When George Zephaniah Church died in London in 1869, the American sons were disappointed when they found their father had left them less money than he had left to their sisters in England and Brazil. A letter from their sister Emma's husband explained the matter. George, Albert, and Adolphus had already received significant funds from their father for buying Grand Prairie farmland and equipment and also had received money from their Uncle Thomas in Southeast Asia. George Zephaniah had not supplied similar funds to his daughters, and he wanted to even things out. "He [their father] intended his daughters to share equally with yourselves," Edward (Ted) Beedell, a solicitor, explained in his 1870 letter to his American brothers-in-law.

But there was a glitch. In his will George Zephaniah had intended to treat his sons and daughters equally, easier to do in England when a landholding was not involved. He was concerned that at his death there might not be enough money to equalize his daughters' inheritance with the money his American sons had already received. So he held some money in a special account for Emma, Adeline, and Alice. As he neared death, however, he realized his estate was large enough for his daughters to receive an inheritance equal to his previous gifts to his sons without tapping the special account.

So he had prepared a codicil to his will, dividing the money in the special account equally among both sons and daughters. But he died before he could sign it.

A Legacy of Gooseberries

Ted, whose wife benefited from the situation, tried valiantly to explain this to George, Albert, and Dolf, but he sounded defensive when he said he couldn't do anything about their losing that part of their inheritance since he is acting as trustee for George Zephaniah's daughters' children. At the turn of the next century, things evened out somewhat when their unmarried sister, Alice, died and bequeathed some of her estate to her American brothers and their families.

Feelings over this matter were eventually repaired, and the letters continued. Once again they expressed fondness and a longing to see one another.

The letters reveal that some time before 1870, Mrs. Bullen, a servant of Elizabeth and George Zephaniah, arrived in the United States. She carried with her a box filled with "a case of instruments" for George W. F., who had displayed artistic talent, and other items from the relatives back in England.

Elizabeth, the men's mother, cherished oil portraits of her long-distance sons Albert and Adolphus, painted when they were about sixteen. After Elizabeth died in 1884, their sister Alice sent Adolphus and Albert their youthful portraits. She wrote to them expressing regret that they had to pay duty on them.

Some of the descendants talked about the "English reunions," and newspaper accounts show they were held regularly into the late 1800s.

The leaders of the Grand Prairie English gradually moved in with one of their children and then died—Sophia Puzey Church in 1874, Emma Dickinson in 1888, Eliza Bentley in 1879, and Richard Puzey Sr. in 1893. Hannah Rymer Puzey died in 1900, but her husband, Henry, lived until 1922, corresponding often and lovingly with his sister

197

Ann, who had been married to Hannah's brother Charles. This Puzey family remained members of the Episcopal church, the American version of the Church of England.

Philip Pusey succumbed in 1890 to the respiratory problems he contracted during the Civil War, and his destitute wife was finally able to get a widow's pension based on doctors' affidavits of his illness. That got her through until her death in 1916.

In 1936 Hattie Clipson was the last of the original immigrants to die.

—ɯ—

As the patriarchs and matriarchs died, they made sure their heirs didn't fight over the property they left them. Typically they treated all their children equally. Over and over the father or mother spelled it out clearly. If any family member formally objected to the will, he or she would get nothing. Pop star Michael Jackson and Brooke Astor, the centenarian philanthropist from New York, also used this "no-contest clause." In Astor's case her son tried to cheat her before she died. He was convicted and sentenced to prison, so no-contest clauses don't always prevent strife.

Some of the Joneses' many descendants still farm the land Henry bought in the 1850s. Puzey and Clipson descendants also continue to own some of the land their English forebears bought. The store Richard and Henry Jones began lasted past the mid-twentieth century. Richard's younger brothers, Frederick and Arthur, ran it for many years and passed it down to Arthur's sons and grandsons.

As close as they were throughout the nineteenth century, the Grand Prairie English must have been focused on the future and not the past. In all the family stories,

A Legacy of Gooseberries

even when there are letters that prove the connections, the intrepid immigrants did not find it remarkable that they had known one another for decades before they set out for an unknown life four thousand miles away. In fact, they barely mentioned it to their descendants. One hundred years later, the connections had been forgotten by all but a few. The gooseberry bushes that lasted into the twentieth century on the farms of the English-born descendants were a hidden clue.

Epilogue

In 1962 a twenty-two-year-old boy and an eighteen-year-old girl, both descendants of the Grand Prairie English, dated a few times and liked one another well enough.

The young people had no idea that their great-great-grandparents had known one another in a city four thousand miles away and had set off on an immigrant adventure together. But both had been told by their parents and grandparents time and again that their land, now divided among many descendants, gave them stature and the makings of wealth, and as yet they had little evidence to the contrary.

The boy asked the girl to marry him. They were about the same ages that their immigrant ancestors had been when they married. The boy did not profess his love for the girl. Instead he told her how he would take care of her. He had a farm, which would be hers also. The land gave them freedom. They would make a good life together. He was an experienced farmer who liked his work, and she suspected he was right when he said they would prosper. He was a lovely boy, gentle and unassuming. His case was tender and compelling. His offer was one his great-grandfather had probably made to his intended and one her great-grandmother would have likely accepted.

Getting to Grand Prairie

But America had changed, and the girl had other dreams. She felt hemmed in by the homogeneity of her small town. She wanted an adventure of discovery, much as her great-great-grandparents had had. She didn't share the belief of most of the Grand Prairie immigrants that land ownership was the best way to autonomy and prosperity. A university education was now a more appealing path to a satisfying life. While the goal was the same, the method of reaching it had changed.

Even in soil-rich Illinois, the land was no longer the only stuff of dreams. She said no.

Maps and Illustrations

English Counties 1851

This map of England shows the counties from which the immigrants came as well as London and other place names mentioned in the book.

Getting to Grand Prairie

James Puzey's house in West Challow is still in good condition and occupied. This is where Richard and Sophia, his children destined for the Grand Prairie, grew up. The archaeologist Stuart Piggott, who inherited the house in 1968 and lived there until his death in 1996, estimated that parts of it were built as early as the seventeenth century. If he is correct, it is the oldest house standing in West Challow today.

Maps and Illustrations

"Fairview," the name Richard Puzey Sr. gave to the area around his farm, had its precursor in West Challow, Berkshire, where Richard grew up. Fairview is located on the green across the road from Richard's boyhood home. It would have had a good view of the frequent fairs and markets that Berkshire's villages enjoyed.

Henry Jones's advertising flyer provides a list of his accomplishments in lighting London. His daughter Eliza took advantage of the gas lights when she attended performances at the theaters he listed.

Maps and Illustrations

Vermilion County

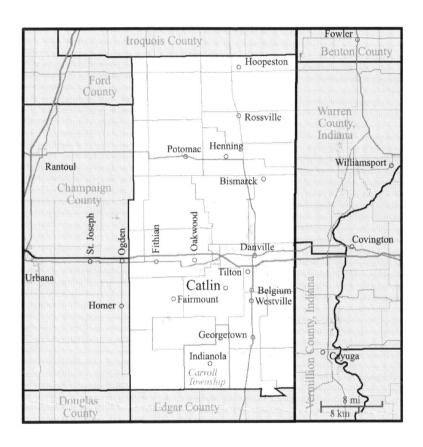

Vermilion County's current boundaries were established in 1859. When the Puzeys, Churches, and Joneses arrived in Vermilion County, there were only two townships. The northern one was Ripley. The southern one was Carroll.

Getting to Grand Prairie

36	31	32	33	34	35	36	31
80 Ch.			6 Miles – 480 Chains			80 Ch.	80 Ch.
1	6	5	4	3	2	1	6
12	7	8	9	10	11	12	7
13	18	17	16	15	14	13	18
24	19	20	21	22	23	24	19
25	30	29	28	27	26	25	30
36	31	32	33	34	35	36	31
1	6	5	4	3	2	1	6

The Public Land Survey System, based on Thomas Jefferson's plan, divided the Midwest and certain other states into rectangles six by six miles square called townships. The townships were further divided into thirty-six equal sections of one square mile each. The sections were numbered beginning at the northeast corner. The numbering continued back and forth until section thirty-six in the southeast corner. The sections could be further divided into halves, quarters and so forth. Such specificity enabled land to be accurately measured for buying and selling.

Maps and Illustrations

Matilda Clipson and Richard Puzey Sr. founded Fairview Methodist Church in the late 1860s. This 1896 building is its second. The first burned.

Getting to Grand Prairie

Richard and Elizabeth Puzey entertained members of the Clipson family in 1903 or 1904. From left to right in the back row are Matilda Puzey Hinton, her husband, Thomas, Albert Clipson, Mrs. Albert Clipson (Ethlen), Richard Clipson, and James Clipson. The two women in front of this row are Mrs. James Clipson (Clarissa Douglas) and Harriett (Hattie) Clipson.

Seated are Richard Puzey (holding Gladys Hinton) and Pauline Clipson, daughter of Albert and Ethlen. To the right are Elizabeth Church Puzey and Russell Clipson, son of Albert and Ethlen.

Seated on the ground are Myrtle Clipson, daughter of James and Clarissa, Lela Clipson, daughter of Albert and Ethlen, and Elsie Puzey, daughter of Richard and Elizabeth.

Many of the Grand Prairie English had died by the twentieth century. In this gathering only James and Richard Clipson and their sister Hattie were original immigrants, coming from London to Vermilion County as children.

A Guide to the Grand Prairie English Families

Family Members and Their Arrivals in America

Date of arrival, ship name, if known: name, age, and identifying phrase.

Note: This information has been gathered from passenger lists, naturalization papers, the U. S. census, Old Settlers' Meeting notes, land purchase records, and family histories.

Getting to Grand Prairie

PUZEY and CHURCH families and their entourages

1847: **Richard Puzey**, 49.

15 July 1848, *Devonshire:* **George W. F. Church**, 18.

1849: **Philip Pusey**, 17, Richard and Sophia's half brother and Phillis's brother.

1849 or earlier: **Henry Church**, 41, husband of Richard's sister, Sophia Puzey.

17 May 1850, *York Town:* **Sophia Puzey Church**, 47, sister of Richard Puzey and wife of Henry Church. Her children with Henry: **Sarah**, 16; **Thomas**, 11; **Jane**, 9.
Albert Church, 18, Henry's nephew.
Phillis Puzey, 16, Sophia and Richard's half sister.
Albert Puzey, 24, Sophia and Richard's nephew, son of their brother Joseph.
Henry Puzey, 26, Sophia and Richard's nephew, son of their brother Joseph.

13 July 1853, *Prince Albert:* Albert Puzey, 26, Sophia and Richard's nephew, son of their brother Joseph, and his new wife, **Eliza Dowding Puzey**, 26, whom he married upon his return to England.
Hester Blanche, 30, Albert's cousin through his mother, Beatrice Blanche.
(An **Alfred Church,** 18, is on this ship. This may be **Adolphus**, 16, nephew of Albert's Aunt Sophia Puzey Church's husband, Henry. Records state that Adolphus

212

arrived in 1853, but he cannot be found on any passenger list. He would not have been the only young man to overstate his age.)

30 May 1853, *Cosmo*: **Frederick Tarrant**, 29, a Puzey cousin.

15 May 1857: **Thomas Puzey**, 19, Henry, Albert, and Jonathan's brother.

1858: **Henry Lloyd**, 17, a Puzey friend or possible relative from Berkshire.

1858: **Hannah Rymer Puzey**, 23, accompanies her new husband, Henry Puzey, when he returns to Illinois.

Late 1850s: **Jonathan Puzey**, brother of Albert, Thomas, and Henry.

Getting to Grand Prairie

JONES and BENTLEY families and their entourage

30 April 1849, *Independence:* **John** and **Sarah Taylor,** both 42. He is a brother-in-law of Henry Shale, Henry Jones's friend.

7 June 1849, *Hendrik Hudson:* **Henry Jones**, 45; **Henry Shale**, 45; **Thomas Hind**, 34.

18 August 1849, *Northumberland:* **Sarah Hough Jones**: 45, Henry's wife, and their children: **Richard**, 25; **Sarah**, 23; **Eliza**, 20; **Emily**, 14; **Louisa**, 8; **Frederick**, 6; **Arthur**, 1.

Ann Hough, 78, Sarah Jones's mother.

Mary Ann Hough, 21, Sarah Jones's niece, daughter of her brother, Samuel Hough.

Elizabeth Hough, 17, Sarah Jones's niece, daughter of her brother, Samuel Hough.

William and **Elizabeth Hind**, both 47. He is the older brother of Henry Jones's friend, Thomas Hind.

Thomas Bentley, 45, and **Eliza Hough Bentley**, 36, Thomas's second wife and Sarah Jones's niece. With them are Thomas's children with his first wife, Sally Holland. They are **Thomas**, 20; **Sarah**, 21; and **Fanny**, 12. Also with them are Thomas and Eliza's children: **Alice**, 5; and **Henry**, infant.

James Bentley, 35, Thomas's brother.

William Browne, 23, Eliza Jones's beloved.

James Thompson, 30, and **Ann Thompson**, 31, and **Ann,** their infant daughter.

30 August 1849, *Danube:* **Edwin Horniblow**, 27, the Joneses' friend mentioned in Eliza's diary.

A Guide to the Grand Prairie English Families

CLIPSON and DICKINSON families and their friends

3 May 1853, *Siddons*: **William Henry Clipson**, 47, and **Matilda Ann Barker Clipson**, 37. With them are the daughters of Will and his first wife, Jane. They are **Catherine**, 18; and **Jane**, 16. Also with them are Will and Matilda's children: **William Henry**, 12; **John**, 9; **James**, 8; **Harriett Ann**, 1½; **Richard**: 4 months.

William Dickinson, 33, and **Emma Barker Dickinson**, 30, Matilda Clipson's sister, and their children: **Harriet Ann**, 5; **Elizabeth**, 3; **William**, 2; **Emma**, 4 months.

Late 1850s: **John Todd**, 23, and **Richard Todd**, 19, William Dickinson's Lincolnshire friends.

1860: **John Carby**, 25, friend from Lincolnshire.

1881: **William**, 59, and **Eliza Carby Dickinson**, 45, John Carby's sister. Their children, **Henry Thomas**, 24; **Stephen Carby**, 22; **Sarah Jane**, 20; **Stephen**, 18; **John A.**, 16; **William**, 14.

Other families

KAY family

Before 1850: **Matthew,** 29, and **Mary Kay,** 28, arrive with **Sarah**, 4. Although Puzey family papers mention the Kays at the English reunions, they can't otherwise be found. A Kay family lived next to the Vandersteens in London.

ONLEYS and ANSTEDS (also spelled OLNEY and ANSTEAD)

6 April 1849, *Westminster:* **Thomas** and **Rachel Onley,** both 44, and their children: **Elizabeth**, 22; **Charles**, 16; **Ann**, 14; **Rachel**, 7; **James**, 9; **Rebecca**, 5; **Benjamin**, 2.

    Charles Ansted, 40, Rachel Onley's brother-in-law, and his children, **Jonathan**, 18; and **Susan**, 14.

1 August 1850, *Garland Grove:* **William**, 26, and **Mary Ann Onley**, 25. William is Thomas and Rachel's son. William and Mary Ann's children: **Matilda**, 2; and **William**, infant.

SWANNELL family

28 October 1847, *Richard Cobden*: **Frederick Swannell**, 22, and **John Swannell**, 24, brothers.

14 September 1848, *Mediator*: **Sarah Swannell**, 44, Frederick and John's widowed stepmother, and her children: **Alfred**, 11; **Henry**, 9; **Eliza**, 7; **Maria**, 5.
Mary Lound, 49, Sarah's sister.
William Swannell, 25, Sarah's stepson.

VANDERSTEEN family

3 November, 1848, *American Eagle*: **George Vandersteen**, 22, and **Jane Burney Vandersteen**, 25, and their son, **George**, 4.
Mary Ann Burney, 55, Jane's mother.
Mary Ann Burney, 22; **Ellen Burney**, 20; and **Sarah Burney**, 19, Jane's sisters.

29 August 1877, *Algeria*: **William Vandersteen**, 66. Widowed brother of George Vandersteen, 52, who returned to England to fetch him.

Selected Bibliography

Ancestry.com. Passenger Lists; U.S. Census Records; English Census Records; Military Records; Parish Birth and Christening Records, England; Illinois Marriage License Records, Agricultural Schedules.

Anderson, Olive. *Suicide in Victorian and Edwardian England.* Oxford: The Clarendon Press, 1987.

Anderson, Ruth Clipson, and Vivian Clipson Coffey. "Descendants of William Henry Clipson" and "Clipson Clippings." Catlin Historical Society, Catlin, Illinois. Privately published, 1977.

Bardsley, Charles Wareing Endell. *A Dictionary of English and Welsh Surnames.* London: H. Frowde, 1901.

Beckwith, H. W. *History of Vermilion County Together with Historic Notes on the Northwest.* Chicago: Knight & Leonard Press, 1879.

Bennett, Stewart, and Nicholas Bennett, eds. *An Historical Atlas of Lincolnshire.* Andover, Hampshire: University of Hull Press, Phillimore & Co., 2001.

Boewe, Charles. *Prairie Albion: an English Settlement in Pioneer Illinois.* Carbondale and Edwardsville: Southern Illinois University Press, 1962.

Bogue, Margaret Beattie. "Patterns from the Sod: Land Use and Tenure in the Grand Prairie, 1850-1900." *Land Series* 1, Springfield: Illinois State Historical Library 34, 1959.

Cameron, Kenneth. *The Place-Names of Lincolnshire.* Nottingham: English Place-Name Society 58, 1980-81.

Carter, Genevieve. "Jones Family History." Catlin Historical Society. Catlin, Illinois: privately published, undated.

Chapman Brothers. *Portrait and Biographical Album, Vermilion County, Illinois.* Chicago: Chapman Brothers, 1889.

Davis, James E. *Frontier Illinois.* Bloomington and Indianapolis: Indiana University Press, 1998.

Dickens, Charles. *David Copperfield.* New York: Signet Classic, 1868-70.

Ditchfield, P. H. *Byways in Berkshire and the Cotswolds.* London: Robert Scott, Roxburghe House, 1920.

Erickson, Charlotte. "Emigration from the British Isles to the U.S.A. in 1841, Part II. Who were the English Emigrants?" *Population Studies* 44, no. 1 (March 1990): 21-40.

——— *Invisible Immigrants: The Adaptation of English and Scottish Immigrants in 19th-Century America.* Ithaca and London: Cornell University Press, 1972.

Faunthorpe, Rev. J. P. *Geography of Lincolnshire.* London: George Philip & Son, 1872.

Gates, Paul Wallace. "Large-scale Farming in Illinois, 1850-1870." *Agricultural History* 6, no. 1 (January, 1932): 14-25.

Gauldie, Enid. *Cruel Habitations: A History of Working Class Housing 1780-1918.* London: George Allen & Unwin, 1974.

Hathaway, Louise G. *The Mann's Chapel Country: A Chronicle of a Pioneer Community, Church and School.* Hoopeston, Illinois: Mann's Chapel Restoration Committee, Mills Publications, 1959.

Hawthorne, Nathaniel. *The English Notebooks.* New York: Modern Language Association of America with the Cooperation of Brown University, 1941.

Hayden, Eleanor G. *Travels Round our Village.* New York: EP Dutton; London: Archibald Constable & Co., 1902.

Houser, Ralph E. "Pusey Family (Direct Line Only) from John 1718 to 1987." Private collection, 1987.

Howse, Violet M., *Pusey, A Parish Record.* Privately published, 1972.

———. *Stanford in the Vale, A Parish Record,* Vol. 1-5. Privately published, 1962.

———. *West Challow, A Parish Record.* Privately published, 1985.

Getting to Grand Prairie

Hunter, Mona Church, and Richard Chrisman. Church Family Papers and Letters. Private collection. Some transcripts are at the Catlin Historical Society, Catlin, Illinois. undated.

Jones Family File. Illiana Genealogical and Historical Society. Danville, Illinois. undated.

Jones, Lottie E. *History of Vermilion County: A Tale of its Evolution, Settlement and Progress for Nearly a Century*, Vol. 1. Chicago: Pioneer Publications, 1911.

Kushner, Howard I. "Immigrant Suicide in the United States: Toward a Psycho-Social History," *Journal of Social History* 18 (Autumn, 1984): 3-24.

Lloyd's Weekly Newspaper. London: British Library Newspaper Collection. 1840s–1850s.

Long, David. *The Little Book of London.* Stroud, Gloucestershire: Sutton Publishing, 2007.

MacKay, Thomas, ed. *The Autobiography of Samuel Smiles, LL.D.* New York: E.P. Dutton, 2005.

Mathews, Milton W. and Lewis A. McLean. *Early History & Pioneers of Champaign County: The Bright and Shady Sides of Pioneer Life.* Urbana: Champaign County Herald, 1891.

McElroy, John. *Andersonville: A Story of Rebel Military Prisons.* Toledo: D. R. Locke, 1879.

Meyer, Douglas K. *Making the Heartland Quilt: A Geographical History of Settlement and Migration in Early-Nineteenth-*

Century Illinois. Carbondale and Edwardsville: Southern Illinois University Press, 2000.

Morning Advertiser. London: British Library Newspaper Collection. 1820-1855.

Nelson, Peter. "A History of Agriculture in Illinois with Special Reference to Types of Farming." PhD diss., University of Illinois, 1931.

O'Brian, Patrick. *Joseph Banks: A Life.* Chicago: University of Chicago Press, 1987.

Parrish, Randall. *Historic Illinois: The Romance of Earlier Days.* Chicago: A.C. McClung & Co., The Lakeside Press, R.R. Donnelly & Sons, 1905.

Peck, J. M. *A Gazetteer of Illinois in Three Parts: Containing a General View of the State, a General View of Each County, and a Particular Description of Each Town, Settlement, Stream, Prairie, Bottom, Bluff, Etc.; Alphabetically Arranged.* Philadelphia: Grigg & Elliot, 1837.

Piggott, Stuart. "The 'Old Hall' in West Challow," Local History Series, Vale and Downland Museum, reproduced from "The Blowing Stone," (Autumn, 1987).

Porter, Roy. *London, a Social History.* London: Hamish Hamilton, Penguin Books, 2000.

Puzey Family File. Illiana Genealogical and Historical Society. Danville, Illinois, undated.

Rawnsley, Willingham Franklin. *Highways and Byways in Lincolnshire*. Illustrated by Frederick L. Griggs. London: Macmillan & Co., 1914.

Stapp, Katherine, and W. I. Bowman. *History Under our Feet: the Story of Vermilion County, Illinois*, Danville, Illinois: Interstate Printers & Publishers and the Vermilion County Museum Society, 1968.

Tilton, G. Wilse [Hannah Mariah, pseud.]. "Catlin Clack," a column in the *Danville Weekly News*. Catlin Historical Society and the Illiana Genealogical and Historical Society. 1880s and 1890s, many undated.

Trollope, Anthony. *The Way We Live Now*. London: Chapman & Hall. 1875.

Urban, Michael A. "An Uninhabited Waste: Transforming the Grand Prairie in Nineteenth-Century Illinois, U.S.A." *Journal of Historical Geography* 31 (October, 2005): 647–665.

Van Vugt, William E. *Britain to America: Mid-19th Century Immigrants to the U.S.* Urbana and Chicago: University of Illinois Press, 1999.

Vincent, James Edward. *Highways and Byways in Berkshire*. Illustrated by Frederick L. Griggs. London: Macmillan & Co., 1906.

Selected Bibliography

White, Jerry. *London in the 19th Century: a Human Awful Wonder of God.* London: Vintage Books, Random House, 2007.

Winn, Christopher. *I Never Knew That About London.* London: Ebury Press, Random House, 2011.

Index

Albert, Prince, 75
Anderson, Joseph, 181
Andersonville Prison, 180, 181
Andrews, Eliza, 158
Anglo-Saxons, 17, 18, 35
Ansted, also Anstead,
    family, 104, 192

Banks, Dorothea, 47
Banks, Sir Joseph, 41-43,
    46, 47, 83, 84
Barker family, 181
Barker, Anne Cumberlidge, 44, 45
Barker, Emma (Dickinson),
    39, 45, 136, 156, 197
Barker, Harriet, 45, 75
Barker, Henry, 45
Barker, James, 44, 45; son 45
Barker, Matilda Anna (Clipson)
    birth, 45; candlestick holder,
    75, 138; church founding, 14,
    187; day of Will's suicide, 3,
    4; death, xiii, 4, 194; fainting,
    181; fresh-air boy, 194; immi-
    gration, 39, 136, 137; land, 140,
    177, 192; marriage, 74; names

for, 45, 74, 75, 79, 133, 193;
    parents, 44, 45, 76, 136; portrait,
    83, 138; receptions, 137
Barker, Thomas, 45
Baum, Charles, 165, 189
Beedell, Edward C. "Ted," 196, 197
Beedell, Emma Maria Church, 64
Belton, Ben, 8
Belton, Dolores "Dee"
    Bedinger and Tom
    ancestors, 7, 12, 13, 14, 15,
    20, 25; Fairview Methodist
    Church, 22; farm 7-11;
    relatives, 30, 116, 119, 188
Belton, Will, 9
Belton, Zach, 8
Bennett, Robert, 189
Bentley family
    arrival, 96; connection to
    other families, 11, 59, 61,
    62, 76, 96, 97, 101, 113, 133,
    168, 178; descendant, 82
Bentley, Alice (Todd, Carby)
    immigration, 113; inheritance
    from Henry Jones, 176, 177;
    marriage, 148, 166, 181, 195

Bentley, Eliza Hough
death, 197; immigration,
113; marriage, 59; mention
in diary, 93, 95, 96; relationship with Henry Jones,
85, 109, 112, 163, 175-177
Bentley, Emily, 96
Bentley, Fanny (Church)
correspondence, 60, 62, 178,
179; immigration, 113; marriage, 62, 148-150; visits, 195
Bentley, Henry, 96, 113,
176, 185, 188
Bentley, James, Thomas's
brother, 60, 85, 113, 163
Bentley, James, Eliza's American-
born son, 176, 177
Bentley, Jessie, 176
Bentley, Sarah Ann (Jones),
60, 113, 148, 179
Bentley, Sarah Holland
"Sally," 60, 62, 85, 178
Bentley, Thomas, father
death, 163, 175; Henry
Jones's will, 177; immi-
gration, 113; land, 125,
126, 128; marriage to
Sarah Holland, 60, 62;
marriage to Eliza Hough,
59; mention in diary; 95,
96; mention in shipboard
letter, 112; occupations,
60, 160; residence, 60, 85

Bentley, Thomas, son, 60,
113, 149, 173, 178, 179
Berkshire
description of, 20-23;
history of, 15-20; hogs, 151;
nonconformists, 100; lamb
with two faces, 116; Puzey
home, 13, 14, 25, 27, 31
betting, 2, 78, 133-136,
150, 169
Birkbeck, Morris, 115, 158
Blanche, Hester, 145
Blanche, Frances "Lizzie," 191
Boston, Lincolnshire, England,
37, 38, 44, 45, 76, 136
Browne, Eliza, 147
Browne, Emily, 147, 169
Browne, William, 93-96,
114, 131, 132, 147
Bullen, Mrs., 197
Burney, Ellen, 101
Burney, Mary Ann, mother, 101;
daughter, 101
Burney, Sarah, 101
Burroughs family, 189, 193
Butler, James, 11
Butler's Point, 11, 116, 119, 124,
125, 128, 156, 160, 161, 164

Cannon, John, 150, 192
Carby, John, 39, 166, 181, 185, 195
Carroll Township, xv, 120, 128, 168
Catlin Clack, 83, 137, 193, 195

Index

Catlin and Catlin Township
benchmark, 160; businesses,
161, 164, 169, 174, 192, 193,
195; Civil War, 172, 173, 180,
181, 182; growth of, 164,
186; land in, 125, 128, 141,
153, 154, 164, 167, 168, 186,
192; leaders, 185; location,
xiv, xv, 116, 120, 127-129,
165, 187, 190; naming and
incorporation of, 161,
178; population, 167, 186,
residents, 169; visitors, 195
Champaign County, 115, 173,
182, 183, 192
Champion, Francis M., 150, 160
Chicago, xiii, 9, 112, 122,
124, 155, 182, 192-194
Childrey, 16, 20, 25; brook, 22
Church family
arrivals, 11; Civil War, 173;
connection to other families,
61-63, 66, 133, 178; deaths of
American-born children,
163; emigration, 84, 85;
letters, 68, 101, 178, 195
Church, Adeline (Pereira), 63, 196
Church, Adolphus Bellingham,
"Doffy, Dolf, and A.B"
citizenship, 165; Civil
War, 178; immigration,
64, 103, 146; land, 168, 196;
marriage, 62, 148-150;
portrait, 197; visits, 195

Church, Albert
connection to other families,
62, 178; immigration, 64, 103,
114; land and investments,
130, 196, 197; leadership, 185;
marriage, 148; portrait, 197;
visits, 195
Church, Alice Augusta,
64, 196, 197
Church, Alfred, 67
Church, Ann Sophia, 67, 68
Church, Charles, 63
Church, Emily Jones, 58,
93, 94, 113, 148, 187
Church, Emma Maria
(Beedell), 64, 196
Church, Edith, 194
Church, Edwina Mary, 194
Church, Elizabeth
Dixon, 62, 63, 66
Church, Elizabeth Lydia
Draper, 63, 64, 67, 197
Church, Elizabeth Sarah
(Puzey), 186
Church, Fanny Bentley
correspondence, 60, 62,
178, 179; immigration, 113;
marriage, 62, 148-150;
visits, 195
Church, George William Frederick
citizenship, 165; connec-
tions with other families,
178, 186; immigration,
64, 103; land, 114, 130, 189,

229

194, 196, 197; leadership, 174; marriages, 132, 194

Church, George Zephaniah, 63, 64, 67, 103, 178, 196, 197

Church, Henry
connections to other families, 61, 146; death, 163; grocery, 67, 68, 160, 161; immigration, 63, 104; land acquisition and sales, 68, 114, 125, 164; marriage, 66; residences, 115, 160

Church, Henry Charles, 67, 68

Church, James, 67

Church, Jane (Champion), 67, 68, 114, 150, 160

Church, Sarah (Lloyd), 67, 68, 114, 166

Church, Sophia Puzey
brothers in London, 64, 65, 67, 68; connection with other families, 61, 63; death, 197; grocery, 65, 68, 160; immigration, 28, 114, 125; marriage, 66; property, 164; relationship with family members, 146, 148, 166; residences, 22, 67, 68, 86, 160

Church, Thomas, father, 62, 63, 66

Church, Thomas, son, 63, 67, 103, 196

Church, Thomas, grandson, 67, 68, 114, 189, 190

Church, William, 63, 103

Clark, Gwen (Taylor), 157

Clipsham, 35, 36

Clipsham, John, 36

Clipson family
Captain Cook, 41; cemetery, 194; connections, 101, 133; Lincolnshire, 33, 35, 41, 46, 74; immigration, 54, 138, 139, 195; inheritance, 44; land, 140, 164, 174, 177, 198; possessions, 75, 83, 138; residences, 1, 44, 48, 72, 75, 94, 133, 136, 139, 141, 142, 150, 173, 192

Clipson, Albert "Tal," 141, 192-194

Clipson, Alexander, 72

Clipson, Bowring, 43

Clipson, Catherine "Carrie" (Moore), 72, 133, 150, 160

Clipson, Charlotta, 43

Clipson, Charlotte Bowring, 39, 43, 44, 76, 136, 138

Clipson, Edwin, 190

Clipson, Ethlen Sanford, 193, 194

Clipson, Frederick, 96

Clipson, Harriet Ann "Hattie," 133, 138, 186, 192, 194, 198

Clipson, James
accent, 82; birth, 79; Civil War, 2, 3, 181; descendants, 141; residences, 133, 193; leadership, 185, 187; marriage, 188

Index

Clipson, Jane Shaw, 41, 48,
69, 72-75
Clipson, Jane (Swannell), 72, 73,
75, 133, 148, 169
Clipson, John, 39, 40, 42,
43, 44, 76, 136
Clipson, John Bowring, 43
Clipson, John Clarence "Jack," 2,
3, 76, 133, 172, 173, 179, 181, 190
Clipson, Lucy Bowring
(Bogg), 43, 136
Clipson, Margaret Hutchings, 190
Clipson, Matilda Anna Barker
birth, 45; candlestick holder,
75, 138; church founding, 14,
187; day of Will's suicide, 3,
4; death, xiii, 4, 194; fainting,
181; fresh-air boy, 194; immigration, 39, 136, 137; land, 140,
177, 192; marriage, 74; names
for, 45, 74, 75, 79, 133, 193;
parents, 44, 45, 76, 136; portrait, 83, 138; receptions, 137
Clipson, Matilda, born 1847,
95; born 1856, 163
Clipson, Marintha Tipton, 190
Clipson, Rebecca (Johnson),
36, 43, 76, 136, 138
Clipson, Richard, 136,
137, 185, 192-194
Clipson, Roy, 141, 142, 180
Clipson, Sarah, first Sarah, 43;
second Sarah, 44, 75, 136

Clipson William Henry "Billie"
birth, 75; Civil War, 2, 3, 172,
179-181; immigration, 133;
Iowa, 190; Prince Albert
story, 77; rickets, 76
Clipson, William Henry "Will"
betting games, 56, 78-80,
133-135; birth, 33, 34, 39, 40;
burial, 194; duel, 142; friendship with Henry Jones, 69,
73, 77, 109, 110, 133, 136, 168;
gas company, 73, 74, 77, 137;
immigration, 136-138; land
purchases, 7, 139, 140, 141,
177, 192; marriages, 41, 45,
74; marriage announce-
ment of daughter Carrie,
150; Metropolitan Police
Force, 70-72; military service,
46-48; parents, 42, 43, 76,
136, 138; portraits, 83, 138;
problems, 135, 136, 138, 141,
142; pubs, 73, 77, 79, 80, 141,
168; residences, 15, 33, 36-38,
69, 72, 73, 75, 76, 79, 133, 135;
suicide, 1-5, 11, 174, 175, 192
Clipson, William John, 48
Clyppsam, Thomas, 35
coal mining, 156, 174, 183, 191
Cook, Captain James, 41, 42
Cork family, 115
Cork, Amelia Jane Eustice
(Puzey), 92, 131, 186, 187, 195

Cork, John, 195
Cork, Reverend William, 91, 131
Crackles, Joseph, Kelsey,
    and Thomas, 47
Craddock family, 102

Danes, 17, 18, 35
Danville, Illinois
    businesses, 155, 156, 160, 179;
    connections with Abraham
    Lincoln, 139, 155, 171; county
    seat, 156; land office, 88, 116,
    124, 128, 164, 182, 186; newspa-
    pers, 137; schools, 12; soldier's
    home, 183; telegraph office,
    136; township creation, 128,
    167; transportation, 112
Davis, Milton, 154
Davis, Jesse, 189
Decatur, Illinois, 191
Derbyshire, 54
Dickens, Charles, 52, 65, 71, 84, 95
Dickinson family, 39, 54, 101,
    138, 139, 142, 143, 181, 186
Dickinson, Emma, 186, 189
Dickinson, Harriet Ann, 186, 189
Dickinson, John and Hannah, 45
Dickinson, John Arthur, 195
Dickinson, William,
    39, 45, 136, 148
Dickinson, William and Eliza
    Carby Dickinson, 195

Dixon, Clementina Bown, 62
Dixon, William, father
    and son, 62
Douglas, Clarissa, (Clipson)
    185, 188, 193

Eastfield Farm, 26
Elbertson, Ellen, 148
Ellsworth, Henry, 139
Ellsworth, Marietta, 139, 140
enclosure, 18-21, 25, 26, 37

Fairchild, Elizabeth, 190
Fairmount, Illinois, 165, 190
Fairview, 13, 22, 23, 116,
    120, 187, 193
Fisher, David, 189
Flower, George, 115, 158

gas light industry and gas
    fitting business, 53-58,
    64, 73-75, 77, 78, 85,
    94-97, 113, 133, 137
George III, 41
George IV, 50, 51, 66
Georgetown, Illinois, xv,
    4, 12, 120, 124, 128, 150,
    160, 173, 192, 194
Gloucestershire, 29, 145, 146, 151
Goings, Serena (Bentley), 188
gooseberries, xvii, 79,
    99, 108, 121, 199

Index

Grand Prairie
attraction of, 83, 84, 90,
102, 103, 106; condition of,
120, 121, 123, 125, 140, 147;
description of, 1, 7, 9-12, 38;
destination, 12, 15, 22, 28-31,
59, 61, 63, 64, 68, 71, 99, 133,
195; farming, 1, 120, 155, 159;
location, xiii, xv, 122, 127;
population, 103, 115, 116;
route to, 111, 112; use of it as
an address, 128, 141, 150, 163
Grant, Ulysses S., 172
Graves, Alice Elisabeth Puzey, 114

Hannah Mariah, 193-195
Hartsough, Walter, 180
Hawthorne, Nathaniel, 36,
44, 46, 58, 76, 79, 88
Hildreth, Alvin, 189
Hind, Thomas, 105-108, 110, 114
Hind, William and
Elizabeth, 112, 113, 125,
160, 174, 189, 194
Hines, George, 194
Hinton, Matilda Puzey, 27, 121
Hommat, Frank, 180
Horniblow, Edwin, 94, 114
Horniblow, Emma, 94
Hough family, 59, 110
Hough, Ann, 54, 113, 130, 147
Hough, Edward, 192, 195

Hough, Eliza (Bentley)
death, 197; immigration,
113; marriage, 59;
mention in diary, 93,
95, 96; relationship
with Henry Jones, 85,
109, 112, 163, 175-177
Hough, Elizabeth, 59, 113
Hough, Mary Ann, 59, 94, 113
Hough, Samuel, father, 54,
59, 111, 113; son, 95
Hough, Sarah (Jones), 54, 58, 59,
67, 85, 93, 96, 107, 113, 163, 175
Hough, William, father,
54; son, 54, 60

Illinois
attraction of, 90, 91, 123, 147,
152, 182; Civil War regiments,
2, 172, 173, 180; farming, 8, 21,
119, 154, 155; governance,
99, 121, 156; history, xiii,
5, 10, 11, 87, 91; land value,
xiv, 87, 126, 202; population, xiii, 49, 87, 89, 121,
122, 123, 152, 167; prairies,
5, 9; roads, 111, 112; surveys,
10, 127; university, 8, 183
immigrants
destinations, 104, 115, 182;
numbers, 100, 104, 122; tensions between, 157-159

233

Indiana, xiii, 9, 11, 83, 87, 89,
  112, 122, 123, 125, 127, 128
inheritance, 44, 140, 176,
  177, 189, 190
Iowa, 87, 155, 190, 194

Jones family
  descendants, xv, 114, 157,
  168, 198; businesses of,
  16, 164, 198; Civil War,
  173; connections to other
  families, 59, 60-62, 73, 97,
  101, 102, 104, 113, 133, 177,
  178; descriptions of, 55, 83;
  homestead, 165; immigration, 11, 53, 59, 85, 93, 96, 100,
  111-114; name origin, 53
Jones, Arthur, 58, 95, 113,
  185, 186, 189, 195, 198
Jones, Eliza (Browne, Tarrant)
  birth, 54; businesses, 161, 192;
  courtship of, 93-97, 114; death
  of William Browne, 147; hot
  cross buns, 186; immigration,
  113, 114; marriages, 131, 132,
  147; move to
  Detroit, 169; restless-
  ness, 148, 168
Jones, Emily (Church), 58,
  93, 94, 113, 148, 187
Jones, Frederick, born
  1841, 55

Jones, Frederick, born 1844,
  58, 113, 114, 130, 164, 169,
  185, 186, 189, 198
Jones Grove Cemetery, 4, 130,
  131, 163, 147, 192, 194
Jones, Henry
  birth, 53, 54; businesses in
  America, 160, 161, 164, 198;
  catching a thief, 57, 58; citizenship, 165; death, 4, 102,
  175, 176; gas fitting, 53-57,
  58, 59, 64, 97; favorite pie,
  xvii, 108; friendship with
  Will Clipson, 4, 69, 73, 77,
  109, 136, 168; immigration,
  59, 105-113; July 4 celebration, 82, 161, 162; land, 125,
  128-130, 139-141, 143, 147, 164,
  198; land value, 154; marriage, 54; shop location, 56,
  57, 96, 133; relationship with
  Eliza Bentley, 85, 96, 109,
  112, 163, 176-178; residences,
  85, 93; will, 101, 164, 176, 177
Jones, Henry, son, 54
Jones, Henry William, 54
Jones, James, 53, 56, 59
Jones, Louisa (Church), 58,
  113, 189, 190, 195
Jones, Rebekah, 54
Jones, Richard, Henry's probable
  brother, 56; Henry's father, 53

Index

Jones, Richard, Henry's son
birth, 54; businesses, 161, 164, 176, 177, 198; catching a thief, 57, 58; death, 186, gas fitting business, 58, 96; immigration, 113; leadership positions, 174, 185; marriage, 148, 179; property, 178, 189; seeing Henry off, 105; troubling letter, 110, 111; trust administrator, 176, 177; wedding host, 148, 149
Jones, Sarah Elizabeth (Church), 54, 113, 132, 186, 194
Jones, Sarah Hough, 54, 58, 59, 67, 85, 93, 96, 107, 113, 163, 175

Kankakee, Illinois, 191
Kansas, 11, 169, 190-192, 194, 195
Kay family, 101, 102
Kirkpatrick, Ann, 28

Lamon, Ward H., 156
land ownership and valuation, 126, 141, 153, 167, 168, 183, 189, 201, 202
Lansdown, Frederick George, 65
Learnard, Mark, 188
Lincoln, Abraham, xiii, xiv, 9, 35, 124, 139, 155, 156, 165, 171, 182
Lincolnshire, 1, 33-40, 39, 45, 47, 148, 166, 195

Lookout Street, 182
London
attractions of, 28, 29, 48, 74, 168; Battersea, 60; Bow, 65; Camden Town, 71; Charing Cross, 64; class division, 81, 82, 102; Clerkenwell, 53; Covent Garden, 56, 57; crime, 57, 58, 70, 80, 86; Dalston, 67; economy, 23, 49; epidemics, 55, 66, 69, 81, 97, 105, 116; growth of, 55; Hackney, 67; jewelry district, 54, 59; Lambeth, 72, 73, 74; modernization, 50-52, 64, 86, 116; newspapers, 65, 66, 69, 90, 95, 150; parish duties, 49, 50, 70; Peckham, 109; population, 49, 85; St. Mary Somerset, 62, 66; Southwark, 69, 76, 79, 80, 104; Stepney, 54, 67, 86; suffrage, 71; Tottenham, 67; voting, 71, 72; Westminster, 66, 69
Lound, Mary Ann, 103
Lloyd family, 20
Lloyd, Henry T., 166, 173, 182, 185, 186, 192

Mann family, 115
Mann, Abraham, 91, 92, 102, 131
Mann, John, 91
McNeer, Valentine, 189

235

Getting to Grand Prairie

Methodists
religion, 12, 38, 73, 100,
156, 188, 192; church, 22,
100, 124, 128, 187, 193
Metropolitan Police Force, 70, 71
Milemore, also Millimore,
family, 102, 176
Miningsby, 33-35, 39, 41,
42, 44, 76, 138
Missouri, 2, 179, 190
Moore, William Mellican, 150
Morris, William, 176

Nash, John, 51
Netherlands, xvii, 34, 37
Newton, Isaac, 38
New York
city, xvii, 53, 96, 105, 110-112,
136, 155, 174; harbor, 101, 103,
104, 109, 112, 113, 115, 125,
136, 138, 151; state, 15, 26, 28,
29, 63, 89, 91, 100, 103, 123
Normans, 15, 18, 35

Oakridge Cemetery, 4, 194
Oklahoma, 190
Onley, also Olney, family,
101, 102, 104, 125, 192
Onley, Benjamin, 172, 173
Onley, Rachel, 139, 140
Onley, Rachel (Cannon),
daughter, 150, 192
Onley, Rebecca (Hough), 192

Onley, Thomas, 125, 129,
139, 163
Onley, William, 115
Ontario, Canada, 15, 26,
28, 83, 100
Oxford University, 16, 27,
35, 36
Oxfordshire, 14, 16, 20, 23, 30, 181

Palmer, Fanny Holland,
62, 178, 179
Parliament
acts of, 19, 37, 50, 71, 72,
135; Clipsham stone,
36; fire, 66, 116, 117
Peel, Sir Robert, 70, 71, 86
Pereira, Francisco de
Paulo Silva, 196
Perry, Robert, 56, 57
prairies
conditions of, 5, 10, 12, 87,
88, 99, 119-122, 126, 155, 169;
names of, 9, 11, 115, 158
Pratt, Capt. Isaiah, 105, 107-109
Presbyterians, 12, 103, 124
Public Land Survey
System, 126, 127, 146
pubs
names of, 29, 56, 76, 79, 80,
94, 141; nature of, 77-79, 80
Puzey, also Pusey, family
burials, 194; coal mining,
174; connections with other

236

Index

families, 11, 61, 62, 66, 101, 102; descendants, 12-14, 18, 114, 157, 188, 198; history of name, 15-17; horn, 17, 18; immigration, 25, 31, 85; grocery orders from England, 160, 186; residences, 20-23, 25, 125, 142
Pusey, Nathan, 31
Puzey, Albert, 30, 31, 114, 130, 145, 168, 190, 191
Puzey, also Pusey, Amanda Jane Anderson, 181, 198
Puzey, Ann (Rymer), 30, 151, 198
Puzey, Ann Tarrant, 25, 26, 29
Puzey, Beatrice Blanche, 29-31, 145, 146
Puzey, Edwin, 146
Puzey, Eliza Dowding, 145, 190, 191
Puzey, Elizabeth Kirkpatrick, 28, 99
Puzey, Elizabeth Sarah Church, 186
Puzey, Frances (Brooks), 30
Puzey, Francis Frederick, 29, 30
Puzey, Frederick, 191
Puzey, George Edwin, 30
Puzey, Hannah Rymer, 151, 197
Puzey, Henry, James and Ann Tarrant Puzey's son, 29, 65, 86

Puzey, Henry, Joseph and Beatrice's son, apprenticeship, 65, 86; argument with Albert, 190; death, 197; education, 29; farm practices, 151; immigration, 30, 114; land, 168, 189, 198; lecture, 87-90, 102; return to England and marriage, 30, 31, 150, 151
Puzey, James, father, 15, 20-22, 25-27, 29, 74
Puzey, James, son, 15, 26, 29
Puzey, Jane, 29
Puzey, John, James's father, 25
Puzey, John, James's son, 15, 26, 29
Puzey, Jonathan, 30, 31, 166, 173, 190-192
Puzey, Joseph, 13, 27, 29-31, 65, 86, 114, 145, 146
Puzey, Mary, 191
Puzey, Mary Seymour, 25
Puzey, also Pusey, Philip, 27, 29, 104, 172, 179, 181, 198
Puzey, also Pusey, Phillis (Holmes), 27, 29, 114, 148
Puzey, Richard, father church founding, 14, 100, 187, 188, 193; death, xiii, 197; early life, 27, 28; Fairview, 22, 120; farm description, xvii, 99, 101, 120-122, 129, 139, 142, 186; immigration, 12, 15, 33, 53, 99; land purchase, 119, 125,

237

143; marriages, 28, 92, 131;
relationship with Sanduskys,
13, 100; visitors, 195
Puzey, Richard, son, 120,
185, 186, 194
Puzey, Roly and Camilla, 30
Puzey, Russell V., 18, 157, 188
Puzey, Sarah Ann "Anna"
(Love), 191
Puzey, Sarah Keen, 26, 27, 29, 74
Puzey, Sarah (Browne), 29
Puzey, Sophia (Church)
brothers in London, 64,
65, 67, 68; connection with
other families, 61, 63; death,
197; grocery, 65, 68, 160;
immigration, 28, 114, 125;
marriage, 66; property,
164; relationship with
family members, 146, 148;
residences, 22, 67, 86, 160
Puzey, Thomas, James and Ann
Tarrant Puzey's son, 29, 65, 86
Puzey, Thomas, Joseph
and Beatrice's son, 30,
166, 173, 185, 191, 192

Quakers, 11, 12, 27, 28

railroads
in America, 89, 111, 112, 115,
154, 155, 160, 161, 163, 164,

182, 183, 192; in England,
37, 44, 67, 73, 134, 116
Romans, 17, 34, 35
Rose King Street, 56,
57, 96, 133, 134
Ross Township and Rossville,
91, 92, 128, 131
Rymer family, 30

St. Louis, Missouri, 59, 111,
113, 122, 124, 192
salt, 10, 124
Sandusky family, 12, 13,
90, 120, 189, 195
Sandusky, Abraham, 13
Sandusky, Isaac, 89, 90, 100,
102, 115, 116, 120, 126, 154
Sandusky, Josiah, 154, 164, 165
Sandusky, Thomas, 165
Sconce, James, 189
Shale, Caroline Taylor, 104
Shale, Henry, 104, 105-110
Shaw, William and Eugenia
Elworthy, 48
Small, Benjamin, 187
Sodowsky, Harvey, 154, 165
South Dakota, 190
Springfield, Illinois, 112, 172, 182
Stanford in the Vale, 17, 19,
20, 23, 25-30, 146, 192
Swannell family, 101, 102,
Swannell, Alfred, 103

Swannell, Eliza, 103
Swannell, Eva, 148
Swannell, Frederick, 101, 103
Swannell, Henry, 103
Swannell, John, 101, 103,
148, 150, 169, 172, 179
Swannell, Maria
Swannell, Sarah Lound, 103
Swannell, William, John's
brother, 103; John's son, 169

Talby, William, 56, 57
Tarrant family, 20, 25
Tarrant, Eliza Jones Browne
birth, 54; businesses, 161, 192;
courtship of, 93-97, 114; death
of William Browne, 147; hot
cross buns, 186; immigra-
tion, 113, 114; marriages, 131,
132, 147; move to Detroit,
169; restlessness, 148, 168
Tarrant, Frederick
businesses, 148, 161, 169,
192; Detroit, 147, 169; immigration, 146; leadership,
185; marriage to Eliza
Jones Browne, 147, 148
Tarrant, Miriam, 169
Tarrant, Sarah, 169
Taylor, John and Sarah,
104, 105, 160, 194
Taylor, Norman, 157

Taylor, Thomas and Ivea,
153, 154, 157, 189
Taylor, Zachary, 121
Thames River, 16, 17, 51-53,
55, 58, 62, 64, 69, 85,
86, 94, 95, 109, 133
Thompson, James, 114,
167, 176, 192
Thompson, Ann, wife, 114, 130
Thompson, Ann, daughter,
114, 130, 147
Tilton, G. Wilse, 137, 138, 193
Todd, John, 39, 46, 148,
166, 173, 176, 181
Todd, Richard, 39, 46, 166, 172, 181

Underground Railroad, 152
Urbana, Illinois, 8, 112, 182, 183

Vandersteen family, 101
Vandersteen, George, father, 101,
176, 185, 194; son, 101, 190
Vandersteen, Jane, 101
Vandersteen, William, 194
Vermilion County
agriculture, 7, 154, 155, 167,
189; Civil War, 154, 172, 173;
description of, 7, 87-90, 115,
122, 131; destination, xiii,
xiv, 13, 101, 111, 115, 116, 131;
English-born population,
xiv, 115, 116, 188, 189, 192;

history, 3, 10-12, 91, 124, 128, 139, 185; land investment or valuation, 126, 141, 153, 183, 189; lecture about, 13, 87, 90, 102, 116, 126; Lincoln leaves, 171; location, xiii, 128; population, 115, 123, 147, 152, 164, 167; transportation, 89, 111, 112, 155, 182

Vermilion River, 10, 122
Victoria, Queen, 66, 75

West Challow, 14, 15, 17, 19, 20-22, 25
William IV, 47, 66
Winch, Isaac, 169
Winthrop, John, 37
Woodbury, W. R., 155

About the Author

Karen Cord Taylor founded *The Beacon Hill Times* in 1995 and later *The Charlestown Bridge* and *The Back Bay Sun* weekly newspapers, serving as the editor and publisher of these entities through 2007. She now writes for those newspapers and other weeklies in her column, "Downtown View." Prior to her newspaper management years, she was a free-lance writer, creating newsletters, corporate materials, and hundreds of newspaper and magazine articles on topics as varied as banking, business, real estate, travel, and design.

She is the author of *Blue Laws, Brahmins, and Breakdown Lanes: An Alphabetic Guide to Boston and Bostonians,* published by the Globe Pequot Press, and, with Doris Cole, *The Lady Architects,* published by Midmarch Arts Press. She was the chair of the committee that published *Hidden Gardens of Beacon Hill: Creating Green Spaces in Urban Places* for the Beacon Hill Garden Club in 2013. Her 2014 book, *Legendary Locals of Beacon Hill,* was published by Arcadia Publishing.

She grew up on a farm in central Illinois and graduated from the University of Illinois. She now lives with her husband, Dan, in downtown Boston.

Made in the USA
Columbia, SC
05 June 2020